THE UNITED KINGDOM PATENT SYSTEM

A BRIEF HISTORY

THE UNITED KINGDOM PATENT SYSTEM

A BRIEF HISTORY

with Bibliography

NEIL DAVENPORT
B.Sc., Ph.D., A.R.C.S., D.I.C.,
Chartered Patent Agent

m KENNETH MASON

Published by Kenneth Mason
Homewell, Havant, Hampshire
© A. N. Davenport 1979

Printed in Great Britain by Coasbyprint Ltd.,
1 Claybank Road, Portsmouth, Hampshire.

ISBN 085937 1 573

CONTENTS

TABLES OF INFORMATION

FOREWORD

Dr Davenport was amongst my earliest course members in the series of Patent Law and Administration lectures that I directed at what subsequently became the City University in the early 1960s.

When he showed me an earlier draft of this book in 1975, I was delighted to find that my own interest in the evolution of the British patent system had communicated itself to at least one kindred spirit in my audience and borne fruit in no small measure. The roles of teacher and pupil had become reversed, thanks to his scholarly industry. I made a few modest suggestions, such as, that he should investigate and report on the role of juries in patent actions, but found it difficult to see how much of what he had written could be improved upon. Nevertheless, he has improved on it and brought it handsomely up to date, taking in his stride the biggest shake-up the system has had in its history.

A legal system is a dynamic thing, constantly reflecting and influencing political, economic and social life, and our patent system is no exception. It is not dabbling in antiquarianism to enquire into the — sometimes quaint — way our system developed. I am unable to share the view of my famous namesake (no relation) that 'history is bunk'. Anyone concerned with the patent system who feels that way now will, I hope, feel differently after perusing the pages that follow. We are all in the author's debt and the way to repay such a debt is to buy him and then read and re-read him.

PETER FORD, London, January 1979

9

PREFACE

On moving from research to patents work in 1962, I attended lectures on patent law and became fascinated by the history of the United Kingdom patent system. I soon had the idea of writing an outline of the subject, separately tracing each main feature from its origin to the present day and giving, where possible, the reasons for the various changes. I was encouraged to believe in the usefulness of such an outline by questions on early patent practice put to me from time to time by Brian Coe, now curator of the Kodak Museum of photography at Harrow. I think that people, like him, who specialise in the history of technical subjects will find this book of most use. To enhance its value as a reference work, I have included tables showing how various details of patent procedure (including fees) have changed over the years, and a full bibliography.

I completed a version of this book in 1975 which Peter Ford, barrister, who was Consultant in Business Law Studies at the City University and has a special knowledge of patents history, most kindly read and criticised. He saved me from a number of errors and suggested improvements which, as far as my sources have allowed, I have included in the present text. I am sincerely grateful to him for his help and encouragement, and for contributing such a generous foreword.

The aims of recent patent law development have been to harmonise the laws of different countries and to introduce systems which simplify, and make somewhat cheaper, the patenting of an invention in several countries. The treaties necessary for establishing two such systems were signed by the United Kingdom in 1970 (Patent Cooperation Treaty) and 1973 (European Patent Convention). Also, a report of a Committee appointed in 1967 'to examine and report with recommendations upon the British patent system and patent law' had appeared in 1970. Thus it was clear by 1975 that a new Patents Act would have to be passed within a few years if the United Kingdom were to be able to ratify these treaties, and participate in the new systems. This is why I decided not to submit my 1975 text for publication, but to wait until the new law and its accompanying rules had appeared, and then to prepare a new text bringing the history up to date. The new law appeared in July 1977; the rules in February 1978.

The Patents Act 1977 has arguably brought about the most fundamental changes in the United Kingdom patent system that have ever been made. Patents are no longer grants made under the royal prerogative by letters patent, and the grounds of invalidity are no longer those developed over centuries by the courts, but are grounds agreed amongst a number of European countries. It will be interesting to see to what extent the Patent Office and the courts will feel justified in referring to decisions made under the old law when they make decisions under the new.

11

Whilst the word 'patent' is associated with inventions in the minds of most people, it is probably at least as familiar from its use in the terms 'patent leather' and 'patent medicine'. I therefore give a brief account of the history of both of these terms even though, as the reader will find, their connections with patents of invention are tenuous.

In collecting information for this book I have consulted early text books, rules and reports held by the British Library, both the Reference Division (once the British Museum Library) and the Science Reference Library (once the Patent Office Library). I wish to thank the staff of those libraries sincerely for their unfailing help.

I also wish to thank Brian Caswell who, when secretary of the 'Patent Office Examining Staff Magazine' committee, allowed me access to some historical articles which appeared in that journal, and my colleague David Lush who read the text in its final stages and made numerous helpful suggestions. I am indebted to Mr. W. Glatzel and his daughter Valerie who, in 1966, arranged for me to borrow a number of early patent documents belonging to Cope and Timmins Ltd of which firm Mr Glatzel was then managing director. I photographed these documents and the dust jacket design is derived from two of the resulting pictures. Sad to relate, the original documents were destroyed some years later in a fire. Lastly, I wish to thank the publishers who not only agreed to publish this book without demur but also took every care to ensure that it was produced exactly as I wished.

INTRODUCTION

Anyone who studies a technical subject soon comes across references to patented inventions and is likely to be helped by some knowledge of patent law and procedure. The aim of this book is to describe in outline how the main features of the United Kingdom patent system originated and developed. The system is of great historical interest because it has the longest continuous history of any. It is not the earliest because the Venetians granted patents regularly from the 1470's onwards, some 80 years before this was done in England. Also the Venetians passed the first patent law, a declaration of the Senate date March 15 1474. However, the Venetian system only continued until the end of the sixteenth century whereas the English system never lapsed and in 1852 formed the basis of a system applying to the whole United Kingdom.

A patent of invention is a monopoly granted by the State which, subject to various conditions, gives the owner the right for a limited time to stop others using the patented invention. It does not, as many people think, give the owner a positive right to use the invention. It could not do so because if the invention were an improvement in one already patented by someone else, it could only be used by agreement with the owner of the earlier patent.

Patents are granted to stimulate industry and do this in two ways: by encouraging people, or firms, to exploit inventions, and causing the details of inventions to be published instead of being kept secret. The encouragement is given by the monopoly which in the case of a good invention can be very valuable. The publication is ensured by requiring an applicant for a patent to file a detailed description of the invention, as part of a 'patent specification', which is later published by the Patent Office. The specification also has to define the scope of the monopoly so that people know what they may or may not do without permission while the patent is in force. It does this by means of at least one 'claim' following the description.

Many people think that it is only necessary to obtain a patent to be assured of a monopoly. This is not so because a patent can only be enforced if it is valid, at least in part. To be valid, it has to satisfy the requirements of the patent law. The most fundamental of these are that the specification must only claim matter which is new and inventive in the light of what was previously known, and must describe the invention well enough to enable others to practise it when the patent has expired.

If a patented invention is used without permission, the owner can bring an action for infringement of his rights. Both parties put forward legal and technical arguments and the court then decides whether what was done was in fact an infringement of any claim and if so whether that claim is valid. If the owner of the patent wins the action, he may be awarded both monetary compensation and an injunction forbidding further infringement. Not surprisingly, the patents which become the subject of infringement actions are usually those of doubtful validity which cover commercially important inventions. 'Strong' patents — those which are almost certainly valid — are normally respected whereas 'weak' patents are likely to be ignored.

The granting of patents for inventions has proved a very effective way of promoting the development of industry, having the great advantage of encouraging innovation at no cost to the State. The owners of patents earn their own rewards and the cost of running a patent office can be met by fees which have to be paid to obtain a patent and keep it in force (the United States patent system is unusual in not requiring the payment of renewal fees to keep a patent in force) and by other sources of income such as the sale of copies of patent specifications. Almost every country now has a patent system and it is interesting to note that Holland, which in 1869 abolished a system dating back to 1817, found it necessary to start a new patent system in 1912 based on a law passed two years earlier.

LETTERS PATENT

The word 'patent' is an abbreviation of the term 'letters patent' derived from the Latin 'litterae patentes' and meaning 'open letters'. It relates to a document issued by the Sovereign, usually addressed to all subjects of the realm, to which the Great Seal of the Realm is attached at the foot so that the document can be read without the seal being broken. Letters patent were used in the Middle Ages, at least since the year 1201, for countless administrative purposes and are still used for a number of these, including the grant of arms and titles of honour, and the appointment of judges. All letters patent were originally prepared in the Chancery, the Office of the Lord Chancellor, the King's secretary and Custodian of the Great Seal.

When monopolies came to be granted for inventions, the grants were made with letters patent. To start with, patents of invention were a minute fraction of all the patents granted, but by the nineteenth century they so greatly outnumbered the others that the word 'patent' gained its present day association with inventions.

Until 1878, all patents were imposing documents engrossed on large sheets of parchment and having the Great Seal in wax (latterly yellow) attached to the foot with a silken cord. In 1878, by the Crown Office Act of 1877, a wafer Great Seal was substituted for the wax one on patents for inventions, and in 1884 the wafer seal of the Patent Office was substituted for the wafer Great Seal and paper was substituted for parchment.

Under the present patent law, patents for inventions are no longer letters patent and a proprietor of a patent is no longer a patentee. Instead of a letters patent document, a person who obtains a patent now receives a certificate of grant.

THE OLD PROCEDURE FOR OBTAINING LETTERS PATENT

In view of the great power of letters patent, elaborate procedures involving checks and counterchecks grew up to ensure that no document was sealed with the Great Seal without the Sovereign's knowledge and express sanction. Much of the procedure in England was laid down by an Act of 1439, a Privy Council Order of 1444, and the 'Clerk's Act' of 1535.

The procedure was as follows. The fees quoted are those which were paid in the 1840's and include the various stamp duties.

1) Petition and declaration.

The inventor prepared a petition to the Crown praying for a patent to be granted to him. He verified the statements in his petition before a Master in Chancery (or, if outside London, before a Master extraordinary). Originally, he did this by swearing an affidavit but after the Statutory Declarations Act, 1835, he made a solemn declaration. The petition included a title describing the invention (Fee 1s 6d = 7½p).

2) Reference to the Law Officer.

The petition and declaration were taken to the Office of the Secretary of State for the Home Department (the Home Office) in Whitehall. Here, a reference of the petition to the Law Officer, written on the petition itself, was signed by the Secretary of State (Fee: £2.2s6d = £2.12½).

3) Law Officer's report.

The signed petition and the declaration were taken to the chambers of either of the two Law Officers, the Attorney-General and the Solicitor-General, the petitioner being allowed to choose which. Here, the Law Officer's clerk compared the title in the petition with the descriptions in any caveats filed in the office by people wishing to be informed of patent applications made in their own technical fields. If there was no apparent conflict of interest, the clerk prepared a report in favour of the petitioner and recommending a period within which a specification describing the invention should be enrolled after a patent had been granted. This report was signed by the Law Officer (Fee: £4.4s = £4.20).

15

4) The King's or Queen's warrant.

The petition, declaration and report were taken to the Home Office where a clerk prepared a warrant authorising the Law Officer to prepare a bill for the intended patent. This was signed by the Sovereign and countersigned by a principal Secretary of State (Fee: £7.13s 6d = £7.67½).

5(a) The Patent Bill.

The warrant was taken to the Patent Bill Office of the Law Officer where is was retained. Here, a clerk prepared on parchment a bill containing the wording of the proposed patent and added at the end a statement ('docket') that the bill had been appropriately prepared. The clerk arranged for the bill and docket to be signed by the Law Officer. He also prepared two transcripts of the bill on parchment and sent one of these to the Signet Office and the other to the Privy Seal Office (Fee: £15.16s0d = £15.80).

(b) The King's or Queen's Bill.

The Patent Bill was taken to the Home Office where it was laid before the Sovereign by the Secretary of State. The sign manual was affixed at the commencement, the document then becoming the King's (or Queen's) Bill (Fee: £7.13s6d = £7.67½). Details of the Bill were entered in a book.

6) The Signet Bill.

The King's (or Queen's) Bill was taken to the Signet Office (in Abingdon Street, Westminster) where a clerk of the signet compared it with the transcript received from the Patent Bill Office. If the documents agreed, the clerk wrote at the head of the transcript a command in the Sovereign's name ordering the Lord Privy Seal to issue a bill to the Lord Chancellor commanding him to make the intended patent. At the foot of the transcript, the clerk testified that the command was issued under the signet and subscribed his name. The signet, a Royal seal held by the Secretary of State in the Home Office, was then affixed to produce the Signet Bill (Fee: £4.7s0d = £4.35). A docket of this bill was entered in a book and the King's (or Queen's) Bill was retained.

7) The Privy Seal Bill.

The Signet Bill was taken by a clerk of the Signet Office to the Privy Seal Office in the same building. There, it was compared with the transcript received from the Patent Bill Office. If the texts agreed, the transcript was headed with a command to the Lord Chancellor that under the Great Seal he should cause letters to be made patent in the form that followed, and completed at the end with the words 'Given under our Privy Seal at our Palace of Westminster', and the date of sealing. The seal was attached with the bill folded (Fee: £4.2s0d = £4.10). Details of the bill were entered in a docket book, and the Signet Bill was filed.

8) The Letters Patent.

The Privy Seal Bill was taken to the Patent Office in Chancery Lane. Here, the seal was cut off and the bill indorsed with the date. The Clerk of the Patents wrote in the margin an official receipt *(recepi)* to be signed by the Lord Chancellor and to include the date which the patent was to bear. The Clerk also engrossed a patent document on parchment copying the wording of the Privy Seal Bill and using the proper conclusion including the words 'By writ of Privy Seal' and his name. He further prepared a docket giving brief details of the proposed grant. The *recepi* and docket were signed by the Lord Chancellor. The patent document and docket were then taken to the sigillator, who, using the docket as his warrant, sealed the patent document with the Great Seal (Fee: £48. 17s0d = £48.85). The documents were taken back to the Patent Office where the letters patent were put in a case and delivered to the patentee (or his agent). The docket was entered into a docket book kept in the Patent Office and accessible to the public.

All patents were enrolled each year in the Enrolment Office, the texts of the Privy Seal Bills being copied onto long narrow strips of parchment, called 'membranes', which were joined together and rolled up.

It normally took six to eight weeks to obtain an English patent by this procedure. The sum of the fees quoted is £94.17s0d (£94.85). If there was more than one inventor this was increased by £18.12s8d (£18.63 approx.) per inventor; if the application was expedited as much as possible, extra fees totalling £5.3s0d (£5.15) had to be paid.

An English patent gave protection in England and Wales, and by payment of additional fees amounting to £4.14s6d (£4.72½) this could be extended to the Channel Islands and any colonies which had made the necessary legislation. An inventor who wanted full United Kingdom patent protection therefore had to obtain separate Scottish and Irish patents, the Scottish through the Chancery Office in Edinburgh and the Irish through the office of the Lord Lieutenant. The respective costs in the middle of the nineteenth century were £82 and £134, and so during that period United Kingdom patenting cost a minimum of £310. If, as likely, the inventor employed an agent to help him through the maze of procedures, the cost was considerably more.

It is not surprising that the cost and complexity of the old patenting

17

procedure attracted comment from Charles Dickens. In his *A poor man's tale of a patent,* he tells how 'Old John' patented his own invention. Dickens's account of the procedure is exact. His hero soon became disenchanted with the whole process as these excerpts show. 'Nobody all through, ever thankful for their money, but all uncivil'. 'Is it reasonable to make a man feel as if, in inventing an ingenious improvement meant to do good, he has done something wrong?' 'Look at the Home Secretary, the Attorney-General, the Patent Office, the Engrossing Clerk, the Lord Chancellor, the Privy Seal, the Clerk of the Patents, the Lord Chancellor's Pursebearer, the Clerk of the Hanaper, the Deputy Clerk of the Hanaper, the Deputy Sealer, and the Deputy Chaff-wax. No man in England could get a Patent for an India-rubber band, or an iron hoop, without feeing all of them. Some of them, over and over again. I went through thirty-five stages. I began with the Queen upon the Throne. I ended with the Deputy Chaff-wax. Note I should like to see the Deputy Chaff-wax. Is it a man, or what is it ?' When Old John parted from his friend Thomas Joy, with whom he had lodged in London for over six weeks whilst obtaining his patent, Thomas said: "John, if the laws of this country were as honest as they ought to be, you would have come to London – registered an exact description and drawing of your invention – paid half-a-crown or so for doing of it – and therein and thereby have got your patent"!

In 1857, five years after the patent system had, at long last, been reformed, Dickens published an interesting and appreciative account of the reformed system in his weekly journal *Household Words.*

THE ROYAL PREROGATIVE AND EARLY PATENTS RELATING TO INDUSTRY

The Royal Prerogative is the power of the Sovereign to rule. It can only be exercised according to precedent and by means of the principal ministers, and it can be curtailed, extended or regulated by Act of Parliament. An ancient prerogative is the privilege of protecting and regulating commerce.

Edward II and Edward III attracted skilled foreign workers to this country by offering letters of protection which overcame the strict guild regulations which forbade competition. The first such grant was made in 1331 to John Kempe of Flanders, a weaver of woollen cloths.

The first grant aimed at introducing a new industry, and so the first English patent for an invention, was made by Henry VI on April 3rd 1449 to John of Utynam for making coloured glass required for the windows of Eton and other college chapels. The term was twenty years. The second English patent, also for a twenty year term, was granted in 1552 by Edward VI to Henry Smyth who intended to introduce foreign workmen expert in making Normandy glass.

MONOPOLIES

The Tudor monarchs found that granting monopolies by letters patent was an effective way of raising revenue. Queen Elizabeth I granted monopolies in everyday necessities including coal, fruit, iron, leather, salt, soap and starch. However, during the last decade of her reign, so much unrest was caused by these monopolies that she issued, in 1601, a proclamation revoking the more obnoxious patents and giving her subjects the right to take cases concerning monopolies to the courts of common law.

This right was exercised almost at once by Edward Darcy, a groom of the Privy Chamber. He had obtained by letters patent a twelve year monopoly in making playing cards. In 1602, he brought an action against Thomas Allin, a haberdasher, in the Court of Queen's Bench for infringing the patent. In a judgement delivered the following year, the court declared the patent invalid.

This famous decision, in what became known as the 'Case of Monopolies', did not, however, end the abuses of monopoly, and James I issued proclamations in 1603, 1610 (the 'Book of Bounty') and 1621 against monopolies. The 'Book of Bounty' declared grants of patents for existing industries to be illegal, and formed the basis of the Statute of Monopolies.

THE STATUTE OF MONOPOLIES, 1623

The Statute of Monopolies, 1623, is a landmark in the history of British patent law because it was the first English statute to refer specifically to patents for inventions. The statute resulted largely from the influence of Sir Edward Coke, a champion of common law, who is today remembered for his famous law commentaries, the 'Institutes'.

The Statute of Monopolies was passed on 25 May 1624 in the 1623-4 Session of Parliament. The title of the statute thus includes the year 1623 because previous to an Act of George III every Act of Parliament was considered as passed and relating back to the first day of the session in which it was passed.

Section 1 rendered all monopolies illegal, but later sections excepted certain monopolies from the general prohibition. Section 6, which is still in force, allowed the grant of monopolies 'for the term of 14 years or under, hereafter to be made of the sole working or making of any manner of new manufactures within this realm to the true and first inventor and inventors of such manufactures, which others at the time of making such letters patent and grants shall not use, so as also they be not contrary to the law nor mischievous to the State by raising prices of commodities at home, or hurt of trade, or generally inconvenient'. As Coke explained, this was largely a declaration of the law as it then existed. The only real change was that the maximum term of a patent was fixed at fourteen years, the time for training two generations of apprentices.

Despite the Statute of Monopolies, the Crown continued to grant objectionable monopolies, and it was not until Parliament gained control of the State finances following the revolution of 1689 that the practice ceased.

THE DEVELOPMENT OF PATENT LAW AND PROCEDURE

The mediaeval procedures concerning letters patent had been evolved for patents of all kinds, and had serious drawbacks for the patenting of inventions. Perhaps the worst of these was the risk an inventor ran of having his invention rendered unpatentable by premature disclosure during the weeks, or even months, which it took him to obtain a patent. The Act of 1439 had forbidden the antedating of letters patent and so an invention was only safe after the Writ of Privy Seal had been delivered to the Lord Chancellor.

Another serious drawback of the old patent system was that legal proceedings for enforcing a patent could seldom take place in a single court. This was because only the Court of Chancery could grant an injunction to stop infringement of patent rights or cancel a patent, whereas only one of the common law courts could determine the validity of a patent or award damages for infringement. Thus, particularly if more than one trial proved necessary, the proceedings could take years and be very expensive.

Although they were much criticised, the cumbersome old procedures remained throughout the Industrial Revolution. A famous critic was James Watt, improver of the steam engine, who in 1790 submitted a paper to the Lord Chancellor making detailed proposals for the reform of the patent system. Societies were formed with the aim of the reform, or in some cases the abolition, of the patent system. There were abortive patent law reform bills in 1819 and 1820, and in 1829 a Parliamentary Committee collected and published evidence on the patent system. However, it did not issue a report making definite proposals for reform.

The first changes in patent law since the Statute of Monopolies were made by Lord Brougham's Act of 1835. This allowed a patentee to amend his patent, an important advantage, but did nothing to simplify or cheapen the procedure for obtaining a patent. Unrest continued to grow, and by 1851 there were no fewer than eleven committees or associations working for the reform of patent law. These included a Select Committee of the House of Lords.

In the end, a private bill of Lord Brougham and a Government Bill were consolidated to produce a Bill which became the Patent Law Amendment Act of 1852. This Act established a single office, The Patent Office, in which all stages of obtaining a patent could be carried out under the control of Commissioners of Patents. The Commissioners were the Lord Chancellor, the Master of the Rolls, the Attorney- and Solicitor-Generals for England, the Lord Advocate, the Solicitor-General for Scotland, and the Attorney-and Solicitor-Generals for Ireland. One patent now gave protection for Great Britain, Ireland, the Channel Islands and the Isle of Man.

The 1852 Act reduced the cost of obtaining a patent, especially where full United Kingdom protection was needed, but offset this reduction by introducing fees which had to be paid to keep the patent in force.

Not all the provisions of the 1852 Act proved satisfactory. A Royal Commission of 1863 under Lord Stanley enquired into the matter and issued a Report in 1865, and a Select Committee of the House of Commons of 1871 made further enquiries and reported in 1872. However, it was not until the Patents, Designs and Trade Marks Act of 1883 that patent law and procedure were again amended extensively. (It may be noted that trade mark law was separated from patent law in 1905 and the law of registered designs was separated in 1949.)

The 1883 Act replaced the Commissioners of Patents by a Comptroller-General of Patents, Designs and Trade Marks serving under the Board of Trade and marks the start of the British patent system as we know it today. The Act greatly reduced the cost of obtaining a patent and almost halved the total renewal fees. It excluded the Channel Islands from the area for which a patent gave protection. This area was again reduced when the Republic of Ireland established its own patent system in 1925.

Since 1883 most of the changes in British patent law have been made in the light of reports by committees as summarized below:

Table 1. Committees appointed to consider patent law

Year of appointment	Chairman	Year of Report	Following Act
1885	Sir Farrer Herschell	1888	1888
1900	Sir Edward Fry	1901	1902
	Lord Parker	1916	1919
1929	Sir Charles Sargent	1931	1932
1944	Sir Kenneth Swan	1945/6/7	1949
1967	Mr (now Sir) Maurice Banks	1970	1977

The most noteworthy changes produced by the Acts listed in the table were as follows. The 1888 Act established a Register of Patent Agents. The 1902 Act provided for patent applications to be examined for the novelty of the inventions concerned, the necessary search being introduced in 1905. The 1919 Act extended the normal maximum term of patent from 14 to 16 years. The 1932 Act specified all the more important grounds on which a patent could be revoked whilst still allowing other grounds to be effective. It also established a Patents Appeal Tribunal, consisting of a specially chosen High Court judge, for hearing appeals against Patent Office decisions, and so took away from the Law Officer his last duties concerning patents. The 1949 Act removed anomalies concerning various patent dates and gave a complete list of grounds of revocation. The 1977 Act introduced the examination of patent applications for obviousness of the inventions concerned, extended the normal maximum term of patents from 16 to 20 years, and made the other amendments necessary for allowing the United Kingdom to participate both in a European patent system and in a system for simplifying the procedure for obtaining patents in more than one country established under the Patent Cooperation Treaty. It also established the Patents Court.

Many of the details of British patent procedure have been determined by rules and orders made on the basis of powers granted by Parliament. Before 1891, many such rules were published, for instance in the *London Gazette* or in Stationery Office publications, but there was no systematic record of them. In 1891, however, all the public and general orders, rules and regulations made in 1890 were collected into an official volume, the first of a series which still continues. The publication of rules was regulated by the Rules Publication Act of 1893, which required all rules to be registered, numbered and, with few exceptions, published. From the 1894 volume onwards, Statutory Rules and Orders were printed with their S.R. & O. Numbers. The 1893 Act was repealed and replaced by the Statutory Instruments Act, 1946 which came into force on January 1 1948, since when rules and orders have been published as numbered series of Statutory Instruments.

In 1970, the Powers of the Board of Trade were transferred to the Secretary of State for Industry, and in 1974 they were transferred from him to the Secretary of State concerned with the Department of Trade.

INTERNATIONAL AGREEMENTS AND THEIR EFFECTS ON UNITED KINGDOM PATENT LAW

Almost every country has a patent system of its own and, at present, an inventor who wishes to protect his invention in a number of countries has to obtain a patent in each. There have been several international agreements for easing the task of patenting in more than one country which have influenced the development of United Kingdom patent law.

The first of these was the International Convention for the Protection of Industrial Property which, as its name indicates, is concerned with all forms of industrial property and so covers trade marks and registrable designs as well as inventions. It was signed in Paris on March 20, 1883 by eleven countries, The United Kingdom was not one of these but acceded to the Convention on March 17 1884. The Convention was ratified by the United Kingdom on June 6, and came into force on July 6, that year. It has been revised at intervals, the latest version being signed at Stockholm in 1967, and preparations for a further revision are planned.

The International Convention forbids discrimination in any Convention country against the nationals of any other Convention country. The most useful provision concerning patents is that which gives an applicant for a patent in a Convention country a period of grace within which he can apply for a patent in any other Convention country and be allowed there the date of his first application. The time originally provided for foreign filings was seven months* but on January 1 1902 this was increased by the 1901 Act to the present twelve month period, following the first revision of the International Convention (Brussels, 1900).

The next fundamentally important international agreement on patents was The Convention on the Unification of Certain Points of Substantive Law on Patents for Invention signed by European countries at Strasbourg on November 27 1963 and hence known as the Strasbourg Convention. This specifies the kinds of invention which can be patented, the conditions for novelty and inventiveness, and the basic requirements for a patent application.

The Strasbourg Convention formed the basis of the Convention on the Grant of European Patents, known more briefly as the European Patent Convention (EPC). This, a 'special agreement within the International Convention', was signed on October 5 1973 in Munich and ratified by the United Kingdom on March 3 1977. The EPC has led to the establishment of a European Patent Office in Munich, in a new building near to the German Patent Office, and also of a branch in Holland, at the Hague, formed by reconstituting the International Patent Institute. This body had been set up by a diplomatic agreement of 1947 to carry out novelty searches for government patent offices and industry and was commonly known as the IIB from the initials of its French name (Institut International des Brevets). The United Kingdom became a member of the IIB in August 1965.

Under the European Patent Convention, it is possible to file at Munich or The Hague or at a national patent office, for instance in London, an application for a European patent designating the countries

*Six months plus an additional month allowed in the case of any country overseas.

in which patents are desired. The application is first considered at the Hague where it is given a preliminary examination in the Receiving Section and is subsequently examined for novelty in a Search Division. It is then published, with the European search report and abstract (if ready), as soon as possible after 18 months from the filing date or, if priority has been claimed from an earlier patent application under the International Convention, from the priority date. Publication earlier still can be requested. The application is later examined by the Examining Division in Munich and, after any necessary amendment and the payment of various fees, results in the grant of an European patent. Despite its name, this is not, in most respects, a single patent covering the designated countries, but is a bundle of national patents, each subject to the law of the country concerned. Thus a European patent (UK) is equivalent to a patent granted by the Patent Office in London.

The next important international agreement to be considered is the Convention for the European Patent for the Common Market, more briefly termed the Community Patent Convention, which was signed at Luxembourg by the nine European Economic Community countries on December 15 1975. When the provisions of this agreement come into force it will be possible to obtain a single patent which covers all the Common Market countries. This will be a true international patent, a single patent effective in a number of countries. Whilst national courts will be able to hear infringement actions, they will not (except as noted below) be able to consider validity, for this will have to be done by the European Patent Office. However, during a transitional period it will be possible in countries like the United Kingdom, where one court can consider both infringement and validity, for the court to declare a Community patent invalid in its own country. Another transitional arrangement is that it will be possible for ten years, or so, to obtain a European patent designating only some of the Common Market countries as an alternative to obtaining a Community patent.

There is one further important international patent agreement: the Patent Cooperation Treaty signed in Washington on June 19 1970. At least 36 countries in Europe and elsewhere have signed or acceded, including the United States of America. The United Kingdom ratified the treaty on October 24 1977 and it came into force on January 24 1978.

The Patent Cooperation Treaty established an International Patent Cooperation Union with the aims, amongst others, of simplifying and rendering more economical the patenting of inventions in several countries, and of assisting the economic development of developing countries. Under the treaty a single application can be made requesting patents in as many participating countries as the applicant chooses to designate. Priority can be claimed from one or more earlier applications. The application is examined formally in the office where it is received and, after any necessary correction, is subjected to a novelty search by an International Searching Authority. The applicant is given one chance to amend his application in the light of the search report. As soon as possible after 18 months following the priority date, the application, if possible with the search report and any amendments, is published.

The application new becomes, in effect, a number of separate patent applications, one in each designated country, and these are prosecuted by the normal procedures of those countries.

The Treaty also provides a system which allows patent applicants of countries which have agreed to be bound by this part (Chapter II) of the Treaty, to have their applications examined by an International Preliminary Examining Authority to obtain 'a preliminary and non-binding opinion on the questions whether the claimed invention appears to be novel, to involve an inventive step (to be non-obvious), and to be industrially applicable'.

The London Patent Office is working for the new international patenting systems in several ways. It acts as a receiving office for both European and Patent Cooperation Treaty (PCT) applications; it is to carry out on behalf of the European Patent Office some substantive examination work on European applications; and, for 15 years, is to act as a Preliminary Examining Authority for PCT applications. It is not an International Searching Authority, novelty searches for PCT Applications filed in London being carrried out by the European Patent Office branch at the Hague.

PATENTABLE INVENTIONS

The present Patents Act (1977) is the first to include a definition of the kinds of invention which can be patented. Earlier Acts, from 1883 on, had defined the word 'invention' as 'any manner of new manufacture the subject of letters patent and grant of privilege within section six of the Statute of Monopolies' and thus, in effect, had said that patents could be granted for inventions of those kinds which the courts had so far considered patentable. However, certain classes of inventions were excluded by successsive Acts, notably inventions of which the use would be illegal or immoral (1883 Act); and frivolous inventions, such as perpetual motion machines, and mixtures of substances intended as foods or medicines having only the properties to be expected from the separate ingredients (1932 Act).

The 1977 Act definition of what is patentable agrees, of necessity, with that of the Strasbourg and European Patent Conventions, and differs little from practice under the 1949 Act. According to this definition, an invention must, to be patentable, be capable of industrial application: that is of being made or used in any kind of industry, including agriculture. Various kinds of subject matter are specifically not allowed, namely discoveries, scientific theories, mathematical methods; aesthetic creations including literary, dramatic, musical and artistic works; schemes, rules or methods for performing mental acts, playing games or doing business; computer programs; ways of presenting information; inventions the publication or exploitation of which would be generally expected to encourage offensive, immoral or anti-social behaviour, inventions not being so regarded, however, only because they are prohibited by law; and any animal or plant variety or any essentially biological process for the production of animals or plants. Microbiological processes and the products of such processes are patentable.

Before the 1919 Act, it was possible — as it is now — to obtain a patent for a new chemical product however prepared. The Act stopped this in order to limit the power of United Kingdom patents obtained by foreign, especially German, chemical manufacturers, and only allowed patents for chemicals when made by particular new processes. This restriction was removed by the 1949 Act.

THE PATENT APPLICANT

The Statute of Monopolies allowed patents to be granted to the 'true and first inventor and inventors' of new manufactures. As early as 1567 the courts had recognised that a person could patent an invention he had seen whilst travelling abroad because the aim of granting patents was primarily to encourage the setting up of new industries here. Thus the term 'true and first inventor' covered either a person who had actually devised a new manufacture or a person who had imported one.

An extension of this principle of invention by importation was that it became possible for a person to apply to patent an invention communicated from abroad. An order of 1859 made it necessary for the name and address of a communicator to be given in a 'communication application', a requirement which was retained until the possibility of making such as application was removed by the 1977 Act.

As made more clear by the 1885 Act, the Act of 1883 allowed the true and first inventor or inventors to apply for a patent jointly with one or more other 'persons'. 'Person' was defined as including a body corporate, such as a company. The Act also amended the law so that the United Kingdom could ratify the International Convention, making it possible for individuals or corporate bodies that had applied for patent protection abroad in a Convention country to make Convention applications for patents here within the allowed period, retaining their priority dates. The 1914 Act (following the 1911 revision of the International Convention) allowed such Convention applications to be made by the assignees or legal representatives of the original applicants.

Although the 1932 Act allowed patents to be granted to assignees of true and first inventors on their own, it was not until the 1949 Act came into force that assignees could apply for patents on their own. To make sure that assignees could not file without the knowledge of the inventors, an assignee applicant had to obtain the assent of each inventor to the making of the application.

Under the present Act (1977) any 'person' (ie individual or corporate body) may make an application for a patent either alone or jointly with another. However, a patent may only be granted to one or more 'persons' having a right to it, each 'person' being an inventor, a 'person' legally entitled to the invention at the time it was made, or a successor in title to an inventor or legally entitled 'person'. Patent applicants have to identify the inventors within 16 months of the earliest priority date and, if they are not themselves the inventors, give details of how they derive the right to apply. If they do not provide this information, their applications are taken to be withdrawn. In the case of an application which does not claim priority from a previous application, inventors who are not applicants are informed of the application by the Patent Office.

THE NAMING OF INVENTORS

Whether or not they are patent proprietors, inventors like to be named in specifications describing their inventions. For non-Convention applications this has never posed a problem because until the 1949 Act, inventors had to be a party to these applications, and so were named as applicants, and after that Act the Patent Office agreed to print the names of inventors at the head of complete specifications regardless of whether or not they were applicants.

For Convention applications the situation has been more complicated because many foreign patent applications are not made by the inventors and, since the 1914 Act, corresponding Convention applications filed in London have not had to be made by the original applicants. Foreign inventors were first given the right to be named in United Kingdom patents by the 1938 Act which implemented revisions of the International Convention made in 1934. Both before and after the 1938 Act, foreign inventors were quite often named by assignee applicants who, as appropriate, described themselves as assignees or communicatees of the inventor in the petition at the beginning of the complete specification.

The 1938 Act gave all inventors, whether in the United Kingdom or abroad, the right to be mentioned in complete specifications, patents and the register of patents, and laid down procedures for correcting errors of omission or superfluity. Equivalent procedures have been provided in subsequent Patents Acts.

The 1949 Act allowed developments of and additions to inventions to be added to basic provisional or foreign specifications when corresponding complete specifications were prepared. To allow for the possibility of the added matter being the work of further inventors, all the inventors had to be named in a form filed before the complete specification was accepted by the Patent Office as being a suitable basis for a patent grant. There is no need for such a late-filed form under the present Act because new matter cannot be added to a specification once it has been filed. Consequently, the details of inventorship which applicants have to supply within 16 months of the priority date cannot later be made incomplete.

It was only in 1966 that the Patent Office agreed to print, as a matter of course, the names of foreign inventors at the head of the complete specifications of Convention applications. The present law gives all inventors the right to be mentioned in published applications and granted patents.

PATENT AGENCY AND THE CHARTERED INSTITUTE

In the early days of the patent system, there were those, including solicitors, who would help inventors obtain patents by the elaborate procedure of the time. Also, some well-known engineers and scientists would help in the drafting of specifications. However, the profession of patent agency seems to have originated in the early nineteenth century. William Hands, author of a patent law text book of 1808, described himself as a 'Solicitor of the Court of Chancery and Agent'. William Newton started to practise as a patent agent in 1820 and in the same year started publication of the *London Journal of Arts* which reviewed the hundred or so patents granted annually. He (in the words of his son Alfred Vincent Newton, also a distinguished patent agent) 'added to the routine work of obtaining the Great Seal, the duties of advising the client as to the novelty and utility of his invention, of describing it in terms that would satisfy the Law Courts, and of securing, when required the invention in foreign countries'. As founder of a firm, Newton and Son, which undertook both United Kingdom and foreign patenting work, William Newton has a strong claim to being the founder of the modern patent agency private practice.

It appears that all those who described themselves as patent agents were not as skilful and honourable as the best, and in 1882, an Institute of Patent Agents was set up to raise standards in the profession. The 1888 Act gave the Board of Trade power to establish a Register of Patent Agents under the care of the Institute, and also forbade anyone to describe himself as a patent agent who was not registered. After an initial period in which any practising agent could register, it became necessary for anyone wishing to become a patent agent to pass examinations set by the Institute. In 1891, the Institute obtained its Royal Charter, and this allowed a patent agent who was a Fellow to describe himself as Chartered Patent Agent (C.P.A.).

Only 70 of the 257 patent agents who registered initially were members of the Institute, and when, in 1893, the Scottish Court of Appeal ruled that the Board of Trade rules of 1889 were *ultra vires* and that a Mr Lockwood could remain on the Register without paying a fee to the Institute, a rival 'Society of Patent Agents' was founded. However the House of Lords overruled the Scottish decision in 1894, and since then the Chartered Institute of Patent Agents has had undisputed control of the profession.

People who had been *bona fide* practising patent agents during the two World Wars were allowed to register without having passed the qualifying examinations by the Acts of 1919 (1920 Rules) and 1949 (1950 Rules).

Since the 1907 Act, the Comptroller has had the power to refuse to recognize as a patent agent a person who has not been on the register.

A person who wishes to carry on for gain in the United Kingdom the business of representing others in respect of matters concerning European patents has to have his name on a list of professional representatives maintained by the European Patent Office. When this list was established in 1978, people qualified to act as patent agents in the contracting states were allowed to have their names entered. Ultimately, entry will be limited to those who pass a European qualifying examination. A United Kingdom professional representative whose name is on the European list may legally describe himself as a European patent attorney.

SPECIFYING THE INVENTION

The early patents were granted to encourage inventors to establish new industries and thereby teach the inventions to the craftsmen and apprentices carrying them out. Only in a few instances was a description of an invention required by the Crown. Perhaps the earliest description was that in Robert Crumpe's patent of April 9 1618 for a method of raising water.

As patents became numerous, it became more necessary to distinguish between them, and this was done either by using a longer title or by inserting a brief description of the invention into the letters patent.

During the reign of Queen Anne, the Law Officer started to insert into letters patent the condition that for the patent to remain valid the patentee had to file a detailed description of the invention in the High Court of Chancery within one month, later six months, of grant. This procedure became standard in 1734. The earliest specification to survive required by the procedure is that filed by John Nasmith for a patent granted on October 3 1711 concerning a method of fermenting wash from sugar and molasses.

Early in the nineteenth century the Law Officer began also to require an inventor to provide a short description of his invention when he applied for a patent so that the invention described in the specification filed after grant could be identified with that for which the patent application had been made. This procedure was made obligatory in 1850 by a Rule of the Attorney-General and was the forerunner of that which applied after the major reform of the patent system brought about by the 1852 Act.

Under the reformed law, a patent applicant was given a choice of application procedures. He could file with his application either a general description of the invention, termed a 'provisional specification', 'describing the nature of the said invention', or a detailed description of the invention, termed a 'complete specification', 'particularly describing and ascertaining the nature of the said invention, and in what manner the same is to be performed'. If he filed a provisional specification he had to file a corresponding complete specification within six months of his patent being granted. Normally, grant had to be within six months of the application date and so by filing a provisional specification, an inventor obtained a period of up to a year in which to develop his invention before filing the complete specification. An inventor who chose to file a complete specification straightaway gained the advantage of having his patent rights starting at the application date.

The 1883 Act required the complete specification to be filed before grant. If a provisional specification was filed with the application, the complete had to be filed within nine months. After 1885 it became possible to extend this time by paying a fee. Since then, the time for filing a 'complete-after' and the possible extension of that time have been as tabulated below. The fee to be paid for extending the filing period varied as shown in Table 7 (page 58).

Table 2 - Time for filing complete specification

Patents Act of	Ref.	Normal time for filing complete specification (months)	Possible extension (months)
1852	13	6	0
1883	34	9	0
1885	37	9	1
1902	50	6	1
1919	62	9	1
1932	69	12	1
1949	88	12	3

A provisional specification only had to describe an invention in sufficient detail to allow it to be identified with that described in the following complete specification. A complete specification served two important purposes. The first was to describe the invention fully so that when the patent had expired, anyone with a working knowledge of the industry concerned who decided to practise the invention had all the information he needed. The second was to define the scope of the monopoly so that, while the patent was in force, people knew exactly what they could and could not do without permission.

Inventors realised by the early nineteenth century that the detailed description needed for the first purpose was not very suitable for the second. They therefore began the practice of adding after the description statements of what they considered to be the features which characterised their inventions. By adding these so-called 'claims' they hoped to obtain broader protection in any actions for infringements of their patents than if they left the scope of their monopolies to be deduced by the court from the title and description.

After the 1883 Act it became obligatory to include at least one claim in a complete specification. However, it was, in effect, held in 1890 (Siddell v. Vickers) that an inventor could still, if he wished, adopt the old established practice of claiming 'the invention substantially as described' and so leave it to the court to determine the scope of his monopoly.

Even after it had become usual for a patent applicant to include detailed claims in his application, it remained common practice for him also to include a claim of the old type. It was felt that it might be possible for the detailed claims to be held invalid and yet for the claim to the particular embodiment of the invention described and illustrated in the specification to be held valid. Such occasions did in fact arise. In a famous case (Raleigh v Miller) concerning a patent for a hub type of cycle dynamo, the only valid claim was that to the illustrated embodiment. A claim of this type is called an 'omnibus claim' because it is a claim 'to everything' (Latin: omnibus, dative plural of omnis) described.

Under the present (1977) Patents Act every specification has to be the subject of a separate application. It is still possible to establish a priority date by filing a specification without claims, the application then being completed by filing claims and an abstract (with the necessary fee) within twelve months. However, if the applicant wishes to expand his specification, he must file a new application with claims and an abstract within the twelve month period, claiming priority from his first application which is then abandoned. Thus apart from having to pay two somewhat larger application fees, a patent application can still, in effect, enjoy the 'provisional' followed by 'complete' specification procedure.

Under earlier Acts it was not possible to make a patent application claiming priority both from a United Kingdom and from a foreign patent application, and yet this could be desirable, for instance where people both here and abroad working for an international company had jointly contributed to an invention. The 1977 Act overcame this and other disadvantages of the previous law by allowing an applicant to claim priority from one or more United Kingdom patent applications and/or from one or more foreign patent applications. All the applications must have been made within the preceding twelve months. However the claims and abstract can be filed later than the other necessary papers, the time for doing so ending a year from the earliest declared priority date, or a month from the filing date, whichever is the later.

DATE AND TERM OF A PATENT

The terms of the earliest patents were individually settled, the first patent of invention having a term of 20 years. The Statute of Monopolies fixed the maximum term at 14 years, and it was not until the 1919 Act, almost three centuries later, that this term was increased to 16 years. The change was made because Britain had been urged at international conferences to provide a term closer to that allowed by other continental countries.

When the present Act came into force, the term was again increased, to 20 years, this being the period agreed under the European Patent Convention. Patents granted under the 1949 Act dated before June 1 1967 ('old existing patents') did not have their term automatically extended to 20 years when the new Act came into force, whereas those dated from then on ('new existing patents') did. It is possible to apply for the term of 'old existing patents' to be extended by up to 4 years.

There are four dates of particular importance relating to patents.

1) The priority date
2) The date at which patent protection starts
3) The date from which legal proceedings can be taken to stop infringement
4) The date at which the patent term starts

Under the old patent system, these dates were all the same, the date on which a patent was granted. A disadvantage of the old system was that an invention might become known during the weeks, or even months, which elapsed between an inventor applying for a patent and the patent being granted. If it did, the patent was invalid because the invention was no longer new. It is therefore not surprising that inventors, or their agents, adopted all kinds of ruses to hasten the grant of their patents and to delay the grant of patents to their rivals.

The 1852 Act abolished the need for speed and secrecy by allowing a patent to have as its date the application date instead of the date of actual grant. Hitherto the ante-dating of all patents had been forbidden by an Act passed in 1439, in the reign of Henry VI.

From the 1852 Act onwards, the four important dates have not all been the same, and their definitions have varied, and so they are considered in turn below.

(1)   The priority date.

As its name suggests the priority date is the date established for an invention when an application is made to patent it. Thus if two inventors independently make the same invention and are both entitled to patent it, it is the inventor who first applies for a patent who is allowed the patent protection. The priority date is important for another reason, because after this date the patent applicant can publish or use his invention without making his subsequent patent invalid.

From 1852 on, a priority date could be established by filing a patent application accompanied by a provisional or complete specification. In 1884 it became possible for an applicant for patent protection abroad to establish a priority date here by filing a patent application at the Patent Office in London under the International Convention within the allowed period (initially seven months but since January 1 1902, 12 months). Until the 1885 Act his priority date was the date of the patent protection he obtained in his own country; since that Act it has been the date when he applied for that protection.

(2)   The date at which patent protection starts.

From the 1852 Act until the present day it has only been possible to obtain damages for infringements carried out after a full patent specification has been made available to the public. Thus under the 1852 Act, damages could only be obtained back to the date of the patent (then, the application date) if the application had been made with a complete specification. If a provisional specification had been filed, damages could only be awarded back to the actual date of grant. From the 1883 Act until when the 1949 Act came into force, the 'complete specification' was published at the acceptance date because the specification was then laid open to public inspection. Under the 1949 Act, the 'complete specification' was no longer made available at acceptance, but some weeks later on an official publication date when printed copies were first sold.

Under the present Act, an application with claims is published soon after 18 months from the earliest priority date, which may be the date of a previous patent application made here or abroad from which priority is claimed or may be the actual application date.

(3) The date from which legal proceedings can be taken to stop infringement

It has never been possible to start infringement proceedings before a patent has actually been granted on a patent application.

(4) The date at which the patent term starts

From the 1852 Act until the time when the 1949 Act came into force, the term of a patent started at its 'date' which, over this period, was the priority date. Part of the term was therefore absorbed by the time which it took to get a patent application accepted. A Convention applicant was worse off because more of the possible patent term was absorbed by the period between his filing the foreign and corresponding British applications. This anomaly was removed by the 1938 Act which made the term of a patent expire 16 years from the date of filing the complete specification.

Other anomalies remained, however. For instance, the acceptance period for a complete specification was reckoned from the application date so that the time for getting the specification in order for acceptance was shorter for a provisional or convention application than for an application filed with a complete specification. Also, the Patent Office novelty search only covered a period up to the application date so that for a provisional or convention application the search did not cover the period between the application date and the date of filing the complete specification. The 1949 Act removed these anomalies by allowing each claim of a complete specification to have its own priority date, and by making the date of filing of the complete specification the key date with regard to the prosecution of a patent application and the maintenance of a patent.

Under the present Act, the 20 year term starts at the date of filing of the application.

THE GROUNDS OF INVALIDITY OF PATENTS

A patent can only be enforced if it is valid or can be made so by amendments which the court is prepared to allow. Hence the question of validity is vital.

Until the 1977 Act, patents were royal grants and a number of fundamental requirements for the validity of such grants had been established under the common law well before the passing of the Statute of Monopolies in 1624. These were that the grant had to be (i) within the law, (ii) not to the prejudice of existing rights, (iii) certain, (iv) not in contradiction of the Sovereign's intention, (v) free from any false consideration or suggestion, and (vi) free from any false recital. Thus, when patent disputes first came to be tried, the decisions were made on the basis of these fundamental requirements, and other requirements arising from the patenting procedure of the day. As the number of decisions grew, and as patent practice developed, various more specific grounds on which a patent of invention could be revoked came to be recognized and were pleaded in patent actions. Although some of these grounds were referred to in earlier Acts, it was only by the 1932 Act that all the important ones were codified. This Act specified 16 grounds of revocation, but still allowed use of any other ground on which it would have been possible, before the 1883 Act, to apply for the repeal of a patent. It was the 1949 Act which first gave a complete list of grounds of revocation. The 1977 Act does likewise, the grounds being those internationally agreed by the Strasbourg and European Patent Conventions.

Up until the 1977 Act, it was possible to oppose an application for the grant of a patent. The grounds on which this could be done were basically grounds on which — had the patent application not been opposed — the validity of the resulting patent could have been attacked. However the grounds of opposition were not worded identically to the grounds of revocation and, like them, varied from Act to Act. Thus the total number of grounds of invalidity that have existed at one time or another is large (about forty). All the more important are considered below in five groups relating to (1) the application, (2) the nature of the invention, (3) the novelty of the invention (4) the specification, (5) changes in how the invention is specified made between application and grant, and (6) the conditions of grant. It is important to recognise that various grounds overlap so that a single defect in, for example, the description of an invention can result in the patent concerned being invalid for more than one reason.

(1)   Grounds of invalidity relating to the application. Patentee not the
      true and first inventor; invention 'obtained'; invention obtained in
      fraud; patent obtained on a false suggestion or representation;
      Convention application made out of time.

The Statute of Monopolies only allowed a monopoly to be granted by letters patent to the 'true and first inventor or inventors' of a manner of new manufacture. Thus if a patent was granted to someone else it was invalid both through being illegal fundamentally and through the person having made a false suggestion in the recitals to the grant by describing himself as the 'true and first inventor'. Tennant's patent of 1798 for a method of making a bleaching liquor was invalidated in 1802 because the court found that a Glasgow chemist had suggested the method to him two years before the patent. A slightly earlier, more famous, but less clear-cut, decision was that of 1785 in which Arkwright's second patent on spinning machinery, dated 1775, was revoked.

He had claimed numerous separate devices including a 'roving machine' for converting a ribbon of carded cotton into a cylindrical strand ready for conversion into thread. The fact that Thomas Hayes had invented this machine in 1767 was one of a variety of grounds on which the patent was revoked.

Under the 1977 Act, a patent can be revoked if it was granted to a person who was not the only person entitled to be granted that patent or to two or more persons who were not the only persons so entitled.

The court has always been ready to help an inventor whose invention has been pirated by means of a fraudulent patent application. For example in 1871, a servant was granted a patent on a provisional specification he had filed when it was found that his master had afterwards obtained, wrongfully, a patent for a similar invention. The master's patent was revoked (Scott v Young).

All the Patents Acts since that of 1852 have provided that where an invention has been stolen (in legal language 'obtained') the rightful owner is not prevented from patenting it himself because of any illicit use or publication. Before the 1977 Act, he could do so by filing a patent application himself and having this accorded the date of the fraudulent application. Under the 1977 Act the Comptroller has power in 'entitlement proceedings' to transfer the rights in a fraudulent application, or a patent granted on one, to the rightful owner. It is only essential for him to file a patent application himself if the fraudulent application has been refused or withdrawn before the proceedings have been concluded, or if the comptroller has ordered the specification of the fraudulent application to be amended so as to exclude the obtained invention.

Another kind of false suggestion in a patent application which until the 1977 Act could cause a resulting patent to be invalid was false claiming of priority under the International Convention. If a Convention application was made more than a year after the first foreign filing, the foreign application referred to in the application not being the earliest, then the patent was invalid (Gumbel's patent). Under the present Act, a wrongful claim to priority causes loss of priority which, if the invention has been published or used after the first application date, can invalidate the patent.

(2) Grounds of invalidity relating to the nature of the invention. Invention not a manner of manufacture; invention not a patentable invention; invention not useful; invention generally inconvenient; invention contrary to law or morality; invention likely to encourage offensive, immoral or anti-social behaviour.

A patent is invalid if it has been granted for a kind of invention which cannot legally be patented. The kinds of invention which may or may not be patented have already been considered in the section 'Patentable

Inventions'. Under the former law an unpatentable invention was 'not a manner of manufacture'. The question of whether or not a particular invention is a suitable subject for a patent is usually settled when a patent application is examined by the Patent Office and so almost all the relevant decisions have been made before grant by the Comptroller, or on appeal from him.

Because patents were first granted for inventions in order to benefit trade, the courts held early on that a patent was only valid if it covered something useful. It was argued that patents for useless inventions would hinder the development of industry. There were some decisions that went further and laid down that to be patentable an invention had to be better than anything previously known. However, these were not followed and, after passing through a period of some uncertainty early this century, the law became settled that an invention was 'useful' if it gave the result promised, explicitly or by implication, in the complete specification. Under the present law, the ground of invalidity 'that the invention is not useful' does not exist.

In the early days of the patent system grant of a patent might be refused for reasons of policy, the invention being — to quote the Statute of Monopolies — 'generally inconvenient'. Thus Coke tells how a patent was not allowed for a method of fulling because it would have enabled as many bonnets and caps to be thickened in one mill in one day as would have needed the effort of fourscore men, and so reduced employment. Probably for the same reason Queen Elizabeth I refused to grant a patent for a frame for knitting stockings invented by William Lee in 1589. However, it was suspected that she was influenced in her decision by the fact that the worsted stockings produced with the frame were so much inferior to the hand-knitted silk hose she wore herself.

The Statute of Monopolies confirmed the common law requirement that a patent grant had to be within the law. Examples of illegal inventions suggested by earlier authors are house-breaking tools (Hindmarch), mantraps (Boehm and Silberston) and a method of forging banknotes (Blanco White and Jacob).

Inventions conducive to immoral behaviour have been refused by the Comptroller, exercising the royal prerogative, under former Acts and the present Act lays down that 'a patent shall not be granted for an invention the publication or exploitation of which would be generally expected to encourage offensive, immoral or anti-social behaviour', such behaviour not being so regarded 'only because it is prohibited by any law in force in the United Kingdom or any part of it'.

(3) Grounds of Invalidity relating to the novelty of the invention. Invention already known; (a) invention prior used in public or in secret; (b) invention prior published; invention prior published in a patent specification or abstract; (c) invention subject of a prior patent application, a valid prior patent or an unpublished prior claim; (d) invention obvious.

37

The Statute of Monopolies confirmed that a patent could legally be granted for 'the sole working or making of any manner of new manufactures within this realm . . . which others at the time of making such letters patent and grants shall not use . . .'. Well before this statute was passed, it had been established that an invention only had to be new in this country to be patentable. This insular definition of novelty, which became inappropriate with the increasing transfer to technical information between countries, was only abolished with the 1977 Patents Act. However, the idea of doing so was by no means new: the Patent Bill of 1851 originally contained a clause requiring such 'absolute novelty'.

In the early years of our patent system, a manner of manufacture was only considered new if it differed significantly from what was already known. Yet it was only in the early 19th century that this concept of 'novelty' was clearly analysed and seen to embrace two kinds of difference from the 'prior art'; a difference in detail and an essential difference which could only result from a display of real ingenuity. In the mid 19th century another term, 'subject matter' came to embody the idea of ingenuity. Originally, this term had its present meaning, the matter with which a patent application or grant is concerned. However, as novelty, in the strict modern sense, came to be distinguished from ingenuity, subject-matter came to mean 'suitable subject matter'. Thus a patent was said 'to have subject matter', if it had been granted for an invention of the right kind which was not obvious in the light of what was previously known. Another term much used in patent law with regard to novelty is 'anticipation'. An invention may be said to be 'anticipated' if its novelty is destroyed by any form of prior knowledge, including prior use, prior publication and prior patenting. However, like 'subject matter', 'anticipation' is a term which has not been used in exactly the same way by all authors and so has to be read (and used) with caution.

The requirement for an invention to be novel was well expressed by Hindmarch in his book of 1846: 'if the public once become possessed of an invention by any means whatever, no subsequent patent can be granted for it, either to the true and first inventor himself, or to any other person, for the public cannot be deprived of the right to use the invention, and a patentee of the invention could not give any consideration to the public for the grant, the public already possessing everything that he could give'.

The novelty requirement of the new Act is that 'an invention shall be taken to be new if it does not form part of the "state of the art",' this comprising 'all matter (whether a product, a process, information about either, or anything else) which has at any time before the priority date of that invention been made available to the public (whether in the United Kingdom or elsewhere) by written or oral description, by use or in any other way' , and also comprising in the case of an invention to which an application for a patent or a patent relates 'matter contained in an application for another patent which was published on or after the priority date of that invention, if . . .

(a) that matter was contained in the application for that other patent both as filed and as published; and
(b) the priority date of that matter is earlier than that of the invention'.

The Comptroller has the power to revoke a patent on his own initiative if it appears to him it was granted for an invention which formed part of the state of the art by virtue of being the subect of an unpublished prior application.

a) Prior use

A patent was held invalid for prior use as early as 1567. Hastings, who had obtained a patent for a method he had imported from Haarlem in Amsterdam for making frisadoes (fine woollen cloths with a long nap) was not able to enforce it because very similar cloths —'baies' — were already made here. Commercial use by an inventor before he obtains a patent has long been held to invalidate his patent. Zink's patent of 1812 for a method of making verdigris was revoked partly because the product had been sold four months before the date of the patent.

The question as to what kinds of secret use before the priority date will invalidate a patent seems never to have been clarified completely. It was early decided that secret use with no commercial exploitation, ie secret experimental use, did not prevent a patent being granted to a later independent inventor. Thus Dollond's patent of 1758 for an achromatic telescope objective was held valid in 1776 even though it was proved that Dr Hall had invented and privately used the same thing in 1720. It was also decided early on that secret commercial use prevented a later patent for the same invention being valid. Tennant's 1798 patent for making bleaching liquor was invalidated in 1802 partly because the defendant proved he had used the method five or six years before the patent date. Where the difficulty lies is in deciding whether if an inventor uses an invention commercially in secret, it being impossible to deduce the invention from the product, he can later on obtain a valid patent for it. The argument against his being able to do so, clearly put in a case of 1837, is that he would effectively obtain a monopoly longer than that granted by the patent. His legal monopoly would not, of course, be extended. It will be interesting to see whether, under the new Act, secret commercial use will be held to form part of the 'state of the art'.

b) Prior publication

One of several reasons why Arkwright's patent of 1775 was held invalid ten years later was that one of the various machines he described in his specification had been published in a book printed (for the third time) in 1773. To be effective, a publication must disclose the invention fully. For example, in 1843 Muntz's patent of 1832 for use of an alloy of specially pure copper and zinc in specified proportions for sheathing ships was upheld even though the specification of a patent of 1800 suggested various alloys, including some of copper and zinc, for the same purpose because the earlier specification described neither the high purity, nor the proportions, specified for the constituents by the later patentee. Where an earlier patent specification has fully described the same invention as a later, then the earlier patent has long been considered a prior publication, it being irrelevant whether the invention had been used or not. However between 1905 and the start of the 1977 Act, United Kingdom (and since 1932 foreign) patent specifications and abridgements over 50 years old were not included in the Patent Office novelty search, and so did not count as prior publications.

Some early decisions indicated that a rare book hard of access might not count as a prior publication but more recently the question has been simply whether the book has been available to the public as of right. This might not be so for a book in a company library, for instance. Under the new law, publication anywhere in the world can destroy novelty. It remains to be seen whether any rules about accessibility will be developed.

c) Prior patenting

An Act of 1515 made it clear that something granted to one person by letters patent could not be granted to someone else by a later patent. Under the old law there was only one date to be considered, the date of grant, in deciding between rival attempts to patent the same invention, and the specification of an earlier patent could be treated as a prior publication. After 1852, however, there were two dates to consider, the application date and the date of grant, and so it became possible to apply to patent an invention similar to one of an earlier application before the specification of that application had been published. Thus the earlier applicant might have a prior (unpublished) claim to the invention claimed by the later. It was then necessary to decide whether the inventions claimed in the rival specifications really were the same. This was difficult because patent specifications are seldom similarly worded, and the efforts made, particularly in recent years, to differentiate rival inventions so that the later applicant was not deprived of a patent led to some controversial decisions (eg Clinical Products patent).

It was to avoid these difficulties that the present law deems the specification of a patent application to have been published at its priority date for the purpose of deciding on the novelty (but not the obviousness) of an invention the subject of a later application. Thus the 'whole contents', not just the claims, of a prior invention application are now considered when the later application is examined by the Patent Office.

d) Obviousness

The Statute of Monopolies included the word 'inventor' but did not refer to 'inventions'. The readiness with which the courts used the word 'inventor' to describe either the deviser of a new manufacture or a person who imported one from abroad suggests that in the 16th and 17th centuries the word did not have its present meaning but meant a person who introduced something new.

Two cases may be cited to show the birth of the concept that mere novelty was not enough to justify the grant of a patent; there had to be an exercise of inventive skill also.

In 1813, Brunton had obtained a patent for 'improvements in the manufacture of ships' anchors, windlasses and chain cables'. As its title suggests, the specification described three inventions. In the first, concerning anchors, the proposal was to form the two arms in one piece having a conical opening in the centre through which the shank was passed. It appeared at the trial held in 1820-1 that, although this method of construction was new for ships' anchors, it had previously been used for adze or mushroom anchors, and was indeed the mode by which the different parts of a common hammer were united together. The patent was therefore invalidated for what was then termed 'lack of novelty', but would nowadays, have been termed 'obviousness'.

In a slightly later case (1838), Losh was unable to enforce his patent of 1830 against Hague because his supposed invention was the construction for use on railways of wheels made in a way which had already been described in an 1808 specification for the construction of wheels for ordinary carriages. In directing the jury, the judge said '. . . it would be a very extraordinary thing to say that because all mankind have been accustomed to eat soup with a spoon, a man could take out a patent because he says you might eat peas with a spoon. The law on this subject is this: that you cannot have a patent for applying a well known thing which might be applied to fifty thousand different purposes, for applying it to an operation which is exactly analogous to what was done before'.

(4)   Grounds of invalidity relating to the specification

a)   The description: insufficiency; best method not described; description unfair or misleading; invention not disclosed clearly or completely enough.

b)   Delineation of the scope of the monopoly: misleading title; old not distinguished from new; claim ambiguous; claim not fairly based; claim covers useless embodiment; claim covers unallowed embodiment.

41

a) The description

It was in a famous case of 1778 – Liardet v Johnson – that Lord Mansfield laid down what is sometimes called 'the doctrine of the specification' : that the main consideration for the validity of a patent is not, as it had previously been, the actual working of the invention, but is the sufficiency of the description of it given in the specification.

Liardet had been granted a patent in 1773 for a cement useful for rendering the surfaces of plain buildings. Three years later he had obtained an Act of Parliament (16 Geo. 3 c. 29) extending the term of this patent by seven years so that it would expire in 1794. The infringer, Johnson, had taken out a patent of his own, and his specification described a composition for the same purpose which did not contain the drying oil and lead specified by Liardet (but contained serum of blood). Expert evidence showed that the drying oil and lead were essential ingredients (the serum being useless) and that they had surreptitiously been added to the compositions actually used by Johnson. In the face of this evidence, Johnson did not attempt to defend his patent.

In summing up for the jury Lord Mansfield said 'the . . point is whether the specification is such as instructs others to make it. For the condition of giving encouragement is this: that you must specify upon record your invention in such a way as shall teach an artist, when your term is out, to make it, and to make it as well as you, by your directions; for then at the end of the term the public have the benefit of it. The inventor has the benefit during the term and the public have the benefit after'. The idea that a patentee was obliged to describe the best way he knew of performing his invention had already formed the basis of a decision by Lord Mansfield, that in which he had invalidated Brand's patent of 1771. When preparing his specification, Brand had deliberately withheld material information (the fact that he used tallow when tempering steel for trusses).

The present Act does not require a patent proprietor to describe his best method, and did not retain another old-established requirement for the specification: that its description of the invention should be neither unfair nor misleading. An early patent revoked for this reason was Savory's of 1815 for making Seidlitz powders. In his specification Savory described at length the preparation of the Rochelle salts, carbonate of soda and tartaric acid required for these powders, suggesting that a laborious process was necessary to the production of any one of the ingredients, when in fact they could all be bought at a chemist's shop.

Under the 1977 Act, 'the specification of an application shall disclose the invention in a manner which is clear enough and complete enough for the invention to be performed by a person skilled in the art'.

b) Delineation of the scope of the monopoly

By the common law, a royal grant was only valid if it was certain, and so letters patent for an invention were only valid if they clearly delineated the scope of the monopoly. This requirement was well expressed in a judgement of 1842: 'a party who obtains a patent is bound clearly to define in his specification what it is he claims to be his invention, in order that the public may know with certainty what they may or may not do without incurring the risk of an action for an infringement of a patent'.

The earliest decisions involving this ground of invalidity were made when it had become obligatory to file, after grant, a specification describing the invention in detail, and so the scope of each monopoly concerned was judged by considering the title and specification together. When, as was often the case, the invention was an improvement in some existing machine or process, the specification had to describe both old and new matter. If the old was not clearly distinguished from the new, the patent could be invalidated, either for claiming something already known or for not defining what was supposed to be new. It was, in part, to avoid invalidity on this ground that inventors started to include claims in their specifications summarising what they considered to be the novel features of their inventions.

As patent law developed, the courts took progressively more heed of the claims when deciding on infringement. If a claim was found to be ambiguous the patent might well be invalidated. For instance, Johnstone's patent of 1844 claimed 'a cable holder to hold without slipping a chain cable of any size'. It could not be decided whether the inventor claimed a cable holder to hold a chain cable of any *one* size, or to hold chain cables of different sizes, and the patent was therefore revoked.

Ambiguity of a claim is not a ground of invalidity under the 1977 Patents Act, although the claim or claims have to be clear and concise.

A patent proprietor is only entitled to a claim of the width needed to protect fully the actual invention made. In a case of 1852-7 concerning a patent for water-raising machinery, the judge said that the law will permit a patentee 'to claim that which he has invented by means of successful experiments or otherwise, and which he has given to the public, but not that which is the mere subject of his speculation or imagination, or of his endeavouring to grasp more than he is entitled to'. In the 1949 Act, this fundamental limitation was the basis of a ground of revocation: 'that any claim of the complete specification is not fairly based on the matter disclosed in the specification'. There is no corresponding ground in the 1977 Act but the Act does require the claim or claims to 'be supported by the description'.

Under the common law, a patent was invalid if it contained any false suggestion. Hence if the invention described in a specification did not produce the promised result, the patent could be revoked. The promise could be in the title, description or claims. A patent invalidated for having a misleading title was that of 1803 to Bainbridge for 'certain improvements in the flageolet or English flute whereby the fingering will be rendered more easy, and notes produced that never were before produced'. The patent was revoked because, although the instrument had been greatly improved, it only produced one new note. Turner's patent of 1781 for a chemical process was invalidated not only because the described process did not produce the product (white lead) specified in the title but also because it described use of 'fossil salt', a term covering several species only one of which, 'sal gem', would answer the purpose.

An example of a patent which was invalidated because it specifically claimed the use of chemicals which did not give the promised result was that of Martin (1834) for making cements and artificial stone. Gypsum, limestone or chalk (calcium carbonate) was mixed with a solution of pearl ash (potassium hydroxide) which was then neutralised with sulphuric acid. The product was dried and heated to red heat. The specification stated 'other acids and alkalis . . . will answer the purposes of my invention . . . ' and the claim was: 'the process of mixing the powdered materials, alkalis and acids as hereinbefore described, and subsequently burning, heating or calcining the same, for the purposes hereinbefore set forth'. At the trial for infringement of the patent, it was proved that nitric acid would not answer for making the cement.

The last described ground of invalidity, the specific claiming of embodiments not giving the promised result, came to be termed 'inutility', the ground in the 1932 and 1949 Acts being 'that the invention is not useful'. There is no corresponding ground in the 1977 Act. This may prove regrettable because the ground discouraged the filing of unduly wide claims.

At times when claims for certain kinds of embodiment of an invention were not allowed, notably claims for chemical substances *per se* which were prohibited between the dates of coming into force of the 1919 and 1949 Acts, the validity of a patent could be attacked on the ground that it contained claims which could not lawfully be made.

(5) Grounds of invalidity relating to changes in how the invention is specified between application and grant. Disconformity with original title: disconformity with provisional specification: new matter invented by another or the subject of an intervening application or old through intervening publication; disconformity with foreign application: new matter the subject of an intervening application or old through intervening publication; matter added after filing; protection extended by unallowable amendment.

It has always been a requirement of United Kingdom patent law that the invention for which a patent is granted must be that for which the patent was originally sought. When only a title was filed with a patent application, the resulting patent could be invalidated if that title was

44

not consistent with the specification subsequently filed. For example, Metcalf's patent of 1816 for 'a tapered hair or head brush' was held invalid the following year because the specification described a brush with bristles which were not tapered but merely of unequal lengths indiscriminately mixed together.

When, after the 1852 Act, a provisional specification could be filed with a patent application, the patent could be invalid if the invention described in the complete specification was different from that described in the provisional. Whilst this ground was considered in a number of cases from the 1850's on, a patent does not seem to have been revoked on it alone until 1882 when Edison's patent of 1878 was held invalid because the complete specification claimed the phonograph in addition to the methods of transmitting sound electrically which had been described in his provisional specification.

A decision as to whether or not there was disconformity in any particular case could be difficult because the purpose of filing a provisional specification was to allow an inventor time in which he could develop his invention before describing it fully in his complete specification. Thus a difference between the inventions described in the two specifications was to be expected. Usually only a fundamental difference (in Edison's case, addition of an altogether new invention) was fatal to validity.

When it became possible under the International Convention to claim priority here from an earlier foreign patent application, disconformity between the foreign and United Kingdom specifications became a ground of invalidity.

The 1907 Act abolished 'disconformity' as a ground of invalidity for the case where matter added to a provisional specification in preparing a complete specification was both new and invented by the applicant. If the added matter happened to form the basis of a patent application made by someone else during the interval between the filing of the provisional and complete specifications, then the maker of the 'intervening' application was given the right to oppose the earlier application. A similar provision was made for the case where an intervening application was filed between a foreign and a corresponding United Kingdom Convention application.

The 1949 Act allowed the claims of a complete specification to have individual priority dates. Thus a claim could either be based on what was described in a preceding provisional or foreign specification, and so be entitled to the date of filing of that earlier specification, or be based on matter first described in the complete specification and so be entitled merely to its own filing date. The validity of each claim could then be decided with reference to its own priority date and the question of disconformity no longer arose. Under the present Act, the priority of inventions, rather than of individual claims, is considered and so different embodiments covered by a single claim are allowed to have different priority dates.

The 1977 Act does not allow matter to be added to a patent specification either during prosecution of a patent application or after grant. It also does not allow the extent of the protection conferred by a patent to be extended by widening of the claims. To add force to these requirements, the Act has introduced two new grounds of revocation which make it possible for a patent to be revoked if either of the forbidden kinds of amendment has been carried out.

(6) Grounds of invalidity relating to the conditions of grant. Invention not put into practice; invention worked exclusively or mainly abroad; objects of compulsory licence not achieved; articles not supplied to government department; invention also protected by a European patent (UK); renewal fees not paid.

Patents granted after 1575 usually contained a clause which invalidated the patent if the invention was not put into practice. This clause was inserted because the original purpose of granting patents for new manufactures was to encourage the setting up of new industries. A proclamation of 1639 declared that patents not put into practice within three years were invalid.

All the Patents Acts from that of 1883 onwards have contained provisions, discussed later under the heading 'Abuse of monopoly', by which a patent proprietor who does not exploit an invention adequately himself can be compelled to grant licences under the patent allowing others to work the invention. Under the Acts of 1902 to 1949, a patent could be revoked if adequate working could not be achieved by such compulsory licensing.

A clause was included in all letters patent granted under the Acts of 1852 to 1949 which rendered the patent void if the patentee refused to supply articles of the invention to a government department.

A patent proprietor is not allowed to protect a given invention simultaneously with a European patent (UK) and a United Kingdom patent. If the Comptroller finds that this is being done, he may revoke the United Kingdom patent.

From the 1852 Act onwards, it has been necessary to pay renewal fees to keep a patent in force for the whole of the possible term. Hence failure to pay the fees has, since then, been a cause for the lapse of a patent.

EXAMINATION OF PATENT APPLICATIONS

Patent applications have always been examined before patents have been granted on them. Until the 1852 Act, examination was by one of the Law Officers. He determined the wording of the patent grant and would object if the title and any description of the invention in the petition were unsuitable. When the specification was filed after grant, he checked that it described the invention for which the patent had been sought. His duties were little changed by the reforms of the 1852 Act. He still checked that the specification — now called a 'complete specification' - described the right invention: that of the title and, if filed after grant, of the 'provisional specification' filed with the application. He additionally checked that the complete specification described only one invention.

The fees from patent applications were the main source of income for the Law Officers. When, during the absence of the Solicitor-General, the Attorney-General discovered that he was receiving far fewer applications than his absent and less-experienced colleague, he insisted on an agreement that they would in future share the applications equally. When the 1852 Act came into force, the Attorney-General received the odd-numbered applications, the Solicitor-General the even. Four guineas (£4.20) was paid by the Treasury for each application examined. This sum was recommended by the Lord Chancellor and the Master of the Rolls, and so two of the Commissioners decided the remuneration of another two! However, the number of patent applications filed increased so much with the reform of the patent system that each English Law Officer received £2,500 in the last quarter of 1852. This rate of income, equivalent to £160,000 per annum in to-day's (1979) money, was thought excessive and the examining fee was immediately halved.

The 1883 Act transferred the examination of patent applications to specially trained Examiners in the Patent Office. They additionally considered whether the inventions were of the right kind for patenting and whether they were sufficiently well described in the complete specification. For a few years the Examiners were further expected, if the title of a new application was similar to that of an earlier, unpublished, application, to compare the two specifications and decide whether they described the same invention. If they did, the Comptroller might then refuse to grant a patent on the later application. Consideration of conflicting applications in this way proved impractical and so the 1888 Act shifted the responsibility, where there was such a conflict, to the later applicant. He was expected within two months of the grant of a patent on the earlier application to relinquish his own right to a patent by abandoning his own application, or if it had already resulted in a patent, by surrendering that.

The examination procedures described so far were designed to ensure that patent applications, and the ensuing patents, were correct in form, but did virtually nothing to increase the likelihood of the patents being valid. For this a literature search is necessary which, according to its thoroughness, will more or less uncover the true relationships between what a patent applicant describes and claims in his specification and what was previously known.

Although a novelty search had been proposed by the Royal Commission of 1863, it was not until after the Fry Committee had shown that 42% of some 900 specifications they arranged to have examined were wholly or partly anticipated, that the 1902 Act provided for a limited novelty search. This was instituted in 1905. The search was restricted to British patent specifications up to 50 years old, because at that time these constituted almost all of the prior specifications. An applicant for a patent was notified of any specifications published before his application date found in the search, but could not be forced to amend his specification in the light of these citations. If he did not do so, however, the Comptroller could insist on references to the relevant citations being inserted in the applicant's specification. The 1907 Act gave the Comptroller the power to refuse an application which claimed the same invention as a prior published specification. It also extended the novelty search to specifications which had been published after the date of the application being examined but which had been filed with earlier applications. It further gave the Comptroller the power to insist on a reference to any such specification if the applicant refused to amend.

Following the recommendations of the Sargent Committee, the members of which had been urged by industry to increase the security and commercial value of patents, the permitted area of the search was extended by the 1932 Act to cover foreign patents specifications up to 50 years old, and other documents of any date, which had been published in this country.

The 1949 Act enabled the Comptroller to insist on a reference to an earlier patent being given in a specification if he considered that it described an invention which could not be performed without there being a substantial risk of that earlier patent being infringed. However, an applicant could avoid having such a reference in his specification if he could show that there were reasonable grounds for contesting the validity of the earlier patent.

The 1949 Act also enabled the Comptroller to consider any documents brought to his notice by an outsider who read a specification when published on acceptance. If the applicant did not amend the specification to the satisfaction of the Comptroller, grant of a patent could be refused. This provision was known colloquially as the 'sneak provision' and appeared to offer a simple and cheap alternative to opposition proceedings. In fact it did not because, since the Comptroller could not consider obviousness during examination and since the 'informer' was not entitled to appear and argue his case, only an exact prior publication was likely to have any effect.

The 1977 Act divided the examination procedure into two stages, the Preliminary Examination and Search, and the Substantive Examination. A fee is payable for each. In the first stage, the application is examined to make sure that it is correct in form and a novelty search is carried out. This search differs in three important respects from that carried out under the Acts of 1932 to 1949. The first is that documents published anywhere, not just those published in the United Kingdom, are relevant. The second is that there is no longer a 50 year limit to the age of patent specifications which can be cited. The third is that patent specifications of earlier priority date published after the priority date of the examined application count as prior publications when the novelty, but not the obviousness, of the invention is considered.

When he has received the search result, the applicant may amend his application which is then published as soon as possible following 18 months from the earliest priority date (which may be the filing date). The published specification includes the abstract, the result of the novelty search, and any new claims filed by the applicant in the light of that search.

Now that the application is published before the whole examination procedure has been carried out, the delay between the filing and publication of a patent application is much less — often by over two years — than it was under the 1949 Act. This is one of the main advantages for industry of the new law.

Within six months of a patent application being published the applicant has to apply for the substantive examination. This differs from the examination carried out under the 1949 Act, the Examiner now considering obviousness as well as novelty. The Swan Committee had recommended this change before the 1949 Act was drafted, but it was made only recently, by the 1977 Act, being necessary to allow the United Kingdom to participate in the international patenting systems agreed by the European Patent Convention and the Patent Cooperation Treaty.

In the substantive examination, the Examiner considers whether the invention is new and inventive in the light of the publications found in the novelty search and of any observations on patentability sent to the Patent Office by outsiders. Now that obviousness can be considered by the Examiner, such observations are likely to be more effective than they were under the 1949 Act. The Examiner then issues a report and the applicant can amend his specification taking into account any objections raised. Terms of six months for a first reply and three months for a subsequent reply are normally set. If the applicant wishes to amend after his first reply, other than in response to the Examiner, he has to pay an additional fee.

A fundamental objection which the Examiner may raise is that the specification describes more than one invention. This question of 'unity of invention' and the related topics of how the contents of patent specifications may be combined or divided are now considered.

UNITY OF INVENTION

Before the 1852 Act it was quite usual for an inventor to save trouble and money by including more than one invention in a single patent. This practice was forbidden by a rule of 1852, and by the 1883 and subsequent Patents Acts. Nevertheless, the law since 1883 has never allowed the validity of a patent to be attacked on the ground that it relates to more than one invention.

Under the present law, the claim or claims of a specification must 'relate to one invention or to a group of inventions which are so linked as to form a single inventive concept'.

DIVISIONAL APPLICATIONS AND COGNATING

Under the 1883 Act it became possible, if an application was refused through containing more than one invention, for the applicant to make further applications to cover the additional inventions. Each new application was given the date of the original one. The 1907 Act allowed a single complete specification to be filed based on more than one provisional specification if these related to inventions which the applicant believed to be cognate or modifications one of the other. If the Comptroller disagreed with the applicant about the interrelationship, he would allow the applicant to divide his complete specification into portions, one for each invention, and prosecute the resulting divisional applications to obtain separate patents, each having the original priority date. A divisional application might also be filed where an application had included in a complete specification both the invention of a provisional specification and a further invention he had made which the Comptroller considered to be separate. This last provision accompanied the abolition of 'disconformity' between complete and provisional specifications as a ground of invalidity.

Somewhat similar provisions about cognating and filing divisional applications were introduced later for Convention applications. The 1932 Act allowed a single complete specification to be based on more than one foreign application made by a given applicant in one country. The 1938 Act no longer required all the foreign applications to have been made by the same applicant: they only had to belong to the applicant at the time he filed his application here. Under these and later Acts, it was possible to divide the complete specification if the Comptroller decided it related to more than one invention.

The 1949 Act abolished 'disconformity' for convention applications, and clarified the claiming of priority, by allowing different claims of a complete specification to have different priority dates. It became possible to claim priority from a number of preceding foreign applications ('multiple priorities'), the inventions being less closely related than was necessary for cognating under the earlier Acts. It also became possible to claim priority from one or more previous foreign applications and the United Kingdom application itself ('partial priorities'). However, it was not until the 1977 Act that it became possible for a specification to derive priority from both United Kingdom and foreign earlier applications.

Under the present law, if the preliminary examination shows that a specification claims more than one invention, only the first claimed invention is subjected to a novelty search. However, the applicant can have a search carried out on any other claimed invention on paying a further search fee. Also, regardless of whether the Patent Office has objected to his application on the ground that it claims more than one invention, he can file a divisional application based on part of the disclosure of the original, 'parent', application.

THE ACCEPTANCE PERIOD

Following the 1883 Act, when patent applications were first examined by the Patent Office, the applicant was allowed a maximum of 12 months from the application date in which to overcome any objections raised during examination so that the complete specification could be accepted. The 1885 Act made it possible to extend this acceptance period by up to three months on payment of a fee. The normal acceptance period and the fee for extending it have varied over the years as shown in the table below. Following the 1949 Act, the acceptance period was reckoned from the date of filing the complete specification instead of from the application date. Under the present (1977) Act an application has to comply with all the requirements within three years and six months of the earliest priority date. This is either the application date or the date of filing of the earliest patent application from which priority is claimed. The period for getting the application in order cannot be extended.

Table 3 Acceptance period

Patents Act or Rule of	Ref	Normal acceptance period (months)	Fee per month extension (£)
1883	34	12 (a)	– (d)
1885	36, 37	12	2.00
1919	62	15	2.00
1932	69	18	2.00
1949	88	12 (b)	2.00
1957	96	42	2.00
1961	106	36	2.00
1964	108, 111	30	2.50
1969	119	30	3.00
1975	134	30	6.00
1975	137	30	8.00
1977	141	42 (c)	– (d)

(a) From application date
(b) From date of filing complete specification
(c) From application or earliest priority date
(d) Extension not possible.

PUBLICATION OF PATENT SPECIFICATIONS

Before 1st January 1849, English patents were enrolled at one of three Chancery Offices at the patentee's choice, namely the Enrolment Office, (for entry on the Close Rolls), the Rolls Office and the Petty Bag Office (for entry on the Specification and Surrender Rolls). From then until the 1852 Act came into force all specifications were enrolled in the Enrolment Office (Acts of 1848 and 1849). The rolls in each of these offices could be inspected (but for many years, no extracts were allowed to be made!). This was so inconvenient that journals, such as the "Repertory of Acts and Manufactures" (1794-1862), were published giving information about new inventions.

After the 1852 Act all specifications were printed, and between 1853 and 1857 the Commissioners of Patents, fired by the enthusiasm of Professor Bennet Woodcroft, the Superintendent of the Specifications, printed virtually all the patents granted since 1617, the date at which the Clerk of Letters Patent in the Court of Chancery started his docket book. A total of 14,359 documents (specifications when these had been filed) were printed, only the 100 or so patents granted before 1617 and the 18 patents granted during the years of the Commonwealth (1649-1660) not being included. Also abridgements of all the printed specifications were published, those for patents published after 1855 containing illustrations.

After 1852 a new sequence of serial numbers for patent applications has been commenced each year. Up till 1916 the patents granted were identified by the application numbers but in that year a new continuous series of numbers starting at 100,001 was used for accepted patent specifications. This number was chosen to avoid confusion with existing application numbers. The numbers of the first published specification of the year for each of the years 1916 to 1979 are given in the following table.

Table 4 Serial numbers of the patent specifications published first in the years 1916 - 79

1916	100,001	1950	634,001
7	102,812	1	648,181
8	112,131	2	663,941
9	121,611	3	685,361
1920	136,852	4	701,181
1	155,801	5	721,191
2	173,241	6	742,701
3	190,732	7	764,681
4	208,731	8	788,351
5	226,571	9	806,871
6	244,801	1960	826,321
7	263,501	1	857,581
8	282,701	2	885,891
9	302,941	3	914,251
1930	323,171	4	945,444
1	340,324	5	978,901
2	364,154	6	1,015,491
3	385,638	7	1,053,401
4	403,718	8	1,097,211
5	421,827	9	1,138,301
6	440,484	1970	1,175,851
7	459,084	1	1,217,901
8	477,516	2	1,258,951
9	498,137	3	1,301,601
1940	516,338	4	1,342,201
1	531,239	5	1,378,941
2	542,237	6	1,419,981
3	550,279	7	1,460,301
4	558,350	8	1,496,751
5	566,451	9	{ 1,537,581
6	574,317		2,000,001
7	583,835		
8	596,286		
9	615,104		

After the 1885 Act, specifications and drawings of applications which had been abandoned or become void were no longer printed or made available to the public. The printing of provisional specifications with the corresponding complete specifications was stopped in 1962, from Specification No. 909,390 onwards, but they can still be inspected at the Patent Office or obtained as photocopies.

Until the end of 1978, each patent application number consisted of a serial number followed by the final two digits of the year. For example, the numbers of the applications filed in 1977 ran from 1/77 to 54422/77. Since January 1 1979 applications have been numbered according to an internationally agreed system in which each number has seven digits, the first two giving the year and the remainder being the serial number. Thus the applications filed in 1979 were numbered from 7900001 upwards.

When the specification of a 1977 Act application is first published, it is given a serial number in a series which started at 2,000,001, to be clear of 1949 Act numbers, and the suffix 'A'. If the specification is republished, on the grant of a patent, the same serial number is used but the suffix is changed to 'B'.

Since March 24 1976 the letter 'B' has been used as a prefix to the number of a printed specification to show that the specification is an amended version of that originally published.

The cost of printed specifications has varied as shown on Table 5. Until 1892 the price depended on the length of the specification, and the table shows the minimum and maximum prices charged.

Table 5. Price of printed specifications

Date	Serial No	Price	
1852	—	2d to 30s 8d	(ca. 1p to £1.53)
March 2 1892	—	8d	(ca. 3p)
·January 1 1915	—	6d	(2½p)
April 13 1916	100,001	6d	(2½p)
April 7 1920	139,446	1s	(5p)
September 7 1948	607,854	2s	(10p)
February 20 1952	666,681	2s 8d	(ca. 13p)
April 6 1955	727,462	3s	(15p)
April 1 1957	771,351	3s 6d	(17½p)
December 6 1961	883,591	4s 6d	(22½p)
January 1 1970	1,175,851	5s	(25p)
January 1 1975	1,380,291	33p	
January 1 1976	1,420,152	75p	
January 1 1977	1,460,301	95p	

Until the 1977 Act, patents of invention were letters patent granted by sealing. Once a patent application was in order for the grant of a patent, the applicant had to request sealing within a specified period. Following the 1885 Act, it became possible to extend this period. The normal sealing periods and possible extensions of these are given in Table 6. If sealing had to be delayed because of opposition or certain other proceedings, different periods applied.

Table 6 Time for sealing

Act or Rule of	Ref	Normal sealing period (months)	Free extension (months) (d)	Maximum additional extension (months)
1852	13	3 (a)	—	—
1883	34	15 (b)	—	—
1885	36	15 (b)	4	—
1907	56	15 (b)	4	3
1919	62	18 (b)	4	3
1932	69	21 (b)	4	3
1949	88	4 (c)	—	3

(a) From the date of sealing of the Warrant for Sealing
(b) From the application date
(c) From the date of publication of the complete specification
(d) Allowed if the time for filing a complete- after-provisional specification or for obtaining acceptance of the complete specification
had been extended.

The sealing fee and the fee for extending the normal sealing period are given in Table 7 (page 58).

The 1949 Act enabled an applicant who had unintentionally allowed his patent application to lapse, through not paying the sealing fee within the extended sealing period, to apply within the following six months for his application to be restored. The fee for applying to restore a lapsed patent application and the fee payable on restoration were the same as the fees given in Table 9B (page 61) for the corresponding stages in the restoration of a lapsed patent. Restoration could be opposed and people who had started to use the invention between the end of the extended sealing period and the date of sealing could continue to do so, subject to certain restrictions. They could not, for instance, install a further machine for the purpose.

Under the present Act, no request for grant additional to that already included in the patent application form has to be made. Therefore a patent is granted as soon as the application is in order. The effective date of grant is the date on which the Comptroller publishes a notice in the Official Journal that the patent has been granted.

THE FEES FOR OBTAINING A PATENT

Before 1852, it had cost a minimum of £310 to obtain full United Kingdom patent protection. Once the English, Scottish and Irish patents had been granted, however, there were no more fees to pay. The Patent Law Amendment Act of 1852 made it possible to obtain a single patent covering the whole United Kingdom for fees totalling only £25, but made it necessary to pay more fees later on to keep the patent in force for the maximum permitted term. Details of these 'renewal fees' are given later.

The 1883 Act further reduced the cost of obtaining a United Kingdom patent by allowing a provisional specification to be filed for £1, and a following complete specification for £3. If a complete specification was filed in the first instance, the cost was £4. When the novelty search was introduced on January 1 1905, it became necessary to pay an additional fee of £1, to pay for the search, when sealing was requested.

The application fee remained unaltered for 95 years, being increased from £1 to £5 by the 1977 Act. The fees for the complete specification, for sealing, for extending the time for filing a 'complete after provisional' specification, and for extending the time for requesting sealing of a patent were all progressively increased as shown in Table 7.

The 1977 Act replaced the single fee paid on filing a 'complete specification' by two equal fees, the first being paid when requesting a preliminary examination and novelty search for a specification with claims, and the second being paid when requesting substantive examination. Although the Act provided for the possibility of a sealing fee being payable, no such fee was required by the 1978 Rules.

Table 7 Fees, in £, payable during the prosecution of a patent application

Act or Rule	Ref	Application fee	First fee	Second fee	Optional extension of time of filing complete	Optional extension of sealing period by 3 months (e)
1883	35	1	3 (a)	—	—	—
1885	37	1	3	—	2.00	—
1904	51	1	3	1 (c)	2.00	—
1907	56	1	3	1	2.00	6.00
1932	71	1	4	1	2.00	6.00
1955	93	1	4	3	3.00	6.00
1961	106	1	10	3	3.00	6.00
1964	111	1	10	3	3.50	7.50
1967	115	1	14	3	3.50	7.50
1969	119	1	14	3	3.50	9.00
1969	121	1	15	6	3.50	9.00
1971	126	1	22	8	5.00	9.00
1974	132	1	25	9	5.00	9.00
1975	134	1	48	17	10.00	18.00
1975	137	1	60	20	13.00	24.00
1978	141	5	40 (b)	40 (d)	—	—

a) Complete specification fee.   b) Search fee.   c) Sealing fee   d) Substantive examination fee.
e) The sealing fee could be extended month by month, the cost per month being one third the figure in this column.

RENEWAL FEES

The Patent Law Amendment Act 1852 reduced the cost of patenting an invention but made it necessary to pay progressively increasing fees to keep the patent in force. The requirement to pay such 'renewal' fees has been a feature of our patent system ever since. It has the advantage of encouraging patent proprietors who find themselves unable to exploit their patents profitably to let them lapse by paying no further renewal fees. The proprietors are thus spared further expense, and the number of existing patents is kept to a minimum.

The renewal fees specified by the 1852 Act were £50, to be paid before the end of the third year of the term of the patent, and £100, to be paid before the end of the seventh. Four fifths of each sum had to be paid as a fee to the Patent Office, the remaining fifth being paid as a stamp duty. However an Act of 1853 substituted stamp duties for the Patent Office fees and from then until March 1964 each patents fee was paid by having the necessary form stamped with an impressed Inland Revenue stamp. Fees paid from March 4 1964 on were paid directly to the Patent Office.

The 1883 Act made the £50 renewal fee payable before the end of the fourth year. In the case of future patents it made the £100 fee payable before the end of the eighth year and offered the alternative of paying all the renewal fees year by year. This method was offered in 1884 to those who had obtained patents under the 1852 Act and had already paid the £50 fee. The annual sums to be paid for the 8th to 14th years of the patent term were 10, 10, 10, 15, 20 and 20 £ respectively.

58

It was by the Patents Rules of 1892 that the present system of having a smoothly increasing scale of renewal fees was started. Table 8 shows how renewal fees have varied from 1883 to the present. The convenient, and at times advantageous, possibility of paying renewal fees for more than one year in advance was removed in 1975.

Table 8 Renewal fees (in £)

RULES OF:	1883	1892	1920	1955	1964	1969	1969	1971	1974	1975	1975	1978
REF NO. :	35	45	66	93	111	119	121	126	132	134	137	141
5	10	5	5	5	6	8	11	13	15	30	40	40
6	10	6	6	6	7	9	12	14	16	32	42	42
7	10	7	7	8	10	12	13	16	18	34	46	46
8	10	8	8	10	12	13	14	18	20	38	50	50
9	15	9	9	12	14	14	16	20	22	42	56	56
10	15	10	10	14	17	17	18	24	26	48	62	62
11	20	11	11	16	20	20	20	26	29	52	68	68
12	20	12	12	17	22	22	22	28	31	60	76	76
13	20	13	13	18	24	24	24	30	34	64	84	84
14	20	14	14	19	26	26	26	34	38	70	92	92
15	–	–	15	20	28	28	28	37	41	76	100	100
16	–	–	16	20	30	30	30	40	45	84	108	108
17	–	–	–	–	–	–	–	–	–	–	–	118
18	–	–	–	–	–	–	–	–	–	–	–	128
19	–	–	–	–	–	–	–	–	–	–	–	140
20	–	–	–	–	–	–	–	–	–	–	–	152
TOTAL £	153	93	126	165	216	223	234	300	335	630	824	1362

(The leftmost label for rows is YEAR OF TERM.)

It can be important for a manufacturer to know whether a particular patent is still in force or whether it has been allowed to lapse through non-payment of a renewal fee. This information can be found from an official register which, for a small fee, can be inspected in the Patent Office. This register has its origin in two registers instituted by the 1852 Act. The first, called the Register of Patents, gave such details as alterations, extensions of term, and expiry; the second, called the Register of Proprietors, gave details of assignments, licences and so on. The two registers were merged by the 1883 Act to give a single Register of Patents which has been maintained ever since.

The Science Reference Library (Holborn Division) keeps a Register of Stages of Progress, based on information published in the Official Journal (Patents), which shows — amongst other things — if a patent has lapsed or expired.

RESTORATION OF A LAPSED PATENT

The 1883 Act enabled a patentee who had accidentally allowed his patent to lapse, through not paying a renewal fee within the proper time, to apply to the Comptroller for an extension of time of up to three months in which to pay. He had to pay an additional fee, according to the extension requested (see Table 9), and to give — and possibly prove — the reasons why he had failed to pay on time. After the three month period, the only way of restoring a lapsed patent was by means of a private Act of Parliament. The Act gave the court power, in the event of an action for infringement of a restored patent, to award no damages for infringements committed between the date of lapse of the patent and the date when the Comptroller ordered the patent to be restored.

The 1907 Act removed the need to show that the failure to pay a renewal fee had been accidental when an application for restoration was made within three months of the date of lapse of the patent. It also made it possible to apply to the Patent Office for the restoration of a patent after three months from its lapse. In this case, proof that the failure to pay had been accidental was required and the application could be opposed by interested parties. Those who had started to use the invention after the patent had lapsed were given a limited right to continue their use under the restored patent.

Table 9 (A) Fees for extending the period in which to pay a renewal fee

Rule of	Ref.	Fee, in £, for extending period by up to:					
		1	2	3	4	5	6 months
1883	35	3	7	10	—	—	—
1892	46	1	3	5	—	—	—
1920	66	2	4	6	—	—	—
1961	104	2	4	6	8	10	12
1964	111	2.50	5	7.50	10	12.50	15
1969	119	3	6	9	12	15	18
1975	134	6	12	18	24	30	36
1975	137	8	16	24	32	40	48

Table 9 (B) Fees for restoring a lapsed patent

Rule of:	Ref	Fee, in £, with application	Fee, in £, on restoration
1907	56	20	—
1949	89	3	10
1968	116	3.50	12
1969	119	4.50	15
1971	126 ·	6	15
1975	134	11	20
1975	137	15	40

The 1949 Act restricted the period within which it was possible to apply for restoration of a lapsed patent to three years from the date of lapse and this period was further reduced, to one year, by the 1977 Act. The 1949 Act also reduced the cost of applying for restoration, but introduced a fee which had to be paid if the application was successful.

When the International Convention was revised in 1958 at Lisbon, it was agreed to allow up to six months for the payment of an overdue renewal fee. The necessary change in the United Kingdom law was made by the 1961 Act.

PATENTS OF ADDITION

The 1907 Act allowed a cheaper kind of patent to be obtained for an invention which was an improvement in or modification of an invention the subject of an earlier patent application or patent. This new kind of patent was called a 'patent of addition' and differed from an ordinary patent in two respects: it did not have to be kept in force by the payment of renewal fees and its term was governed by that of the patent for the main invention.

The 1949 Act enhanced the probable validity, and so the value, of a patent of addition by forbidding any attack on its validity being made on the ground that it covered an invention obvious in the light of the main invention or on the ground that it covered an invention which should have been the subject of an independent patent.

Patents of addition are not obtainable under the 1977 Act. The main reason for their abolition is that applications for patents of addition are just as expensive to process in the Patent Office as applications for ordinary patents.

LICENCES OF RIGHT

The 1919 and subsequent Acts have enabled a patent proprietor who is willing to licence anyone under his patent to have the patent indorsed 'licences of right' and thereafter pay only half the normal renewal fees. Disputes as to the terms of a licence can be referred to the Comptroller.

CHANGES IN THE COST OF PATENTING

It is interesting to work out how the cost of obtaining and maintaining patent protection in the United Kingdom has changed, in real terms, over the years. This has been done by using the information on fees already presented in conjunction with figures for the relative purchasing power of the pound based on cost of living data. The costs of obtaining a patent are taken largely from Table 7; those for maintaining a patent from Table 8. No attempt has been made to include the cost of the services of a patent agent. Few data have been published, and in any case the cost has always varied with the length and difficulty of the specifications involved.

Table 10   Changes in the cost of United Kingdom patenting

Year	Purchasing Power of £1 relative to Feb 1978	Cost of obtaining patent		Cost of maintaining patent	
		Actual £	Relative £	Actual £	Relative £
1829	15.4	310	4770	0	0
1853	16.1	25	403	150	2420
1883	15.5	4	62	150	2330
1892	18.3	4	73	95	1740
1920	7.05	5	35	126 (a)	888
1932	17.2	6	103	126	2170
1950	5.79	6	35	126	730
1955	4.45	8	36	165	734
1964	3.43	14	48	216	741
1969	2.77	22	61	234	648
1971	2.39	31	74	300	717
1974 (1st qtr)	1.89	35	66	335	633
1975 (1sr qtr)	1.57	66	104	630	989
1975 (3rd qtr)	1.37	81	111	824	1130
1978 (Feb)	1.00	85	85	1362 (b)	1362

(a) Term increased from 14 to 16 years
(b) Term increased from 16 to 20 years

EXTENSION OF TERM

Up till 1835, a private Act of Parliament was required to extend the term of a patent on the ground that the patentee had not been able to earn adequate reward. A famous instance was the extension to a term of 25 years which James Watt obtained by an Act 1775 for his patent of 1769 for improvements in the steam engine.

The 1835 Act enabled the Crown to extend a patent for seven years if the Judicial Committee of the Privy Council recommended this and the 1844 Act made a further seven years extension possible in exceptional cases. The power to decide on extensions of term was transferred to the High Court by the 1907 Act. The 1919 Act reduced the maximum possible extensions to 5 and 10 years and also allowed an extension to be obtained on the ground that war had caused the patentee to suffer loss or damage. After 1946 it became possible to apply to the Comptroller, as well as to the High Court, for a war loss extension.

No extension of the term of patents granted under the present Act is possible.

OPPOSITION TO GRANT

From the early days of the British patent system until June 1 1978, when the 1977 Act came into force, it was possible to oppose the grant of a patent. By the 17th century, it had become usual for a person having an active interest in a particular industry to enter a caveat in one of the offices through which petitions for patents were passed requesting to be informed of any petition relating to an invention in that industry. A caveat was for three months, later a year, but could be renewed on payment of a further fee. On being informed of a petition, the person who had entered the caveat could lodge a notice of opposition within a week. If he did so, he and the petitioner would be heard separately by the Law Officer to prevent disclosure of the invention to the person opposing. The Law Officer would decide on whether the petitioner would obtain his patent. Since the opponent was working in the dark, an opposition was seldom successful. If it was, the petitioner had no chance to appeal.

Under the 1852 Act, an applicant for a patent gave notice of his intention to proceed with his application in the *London Gazette* (after 1878 in the *Commissioners of Patents' Journal*). An interested party could oppose within three weeks, and the matter was referred to the Law Officer. It was possible to appeal against the Law Officer's decision to the Lord Chancellor.

It was the 1883 Act which introduced, in general form, an opposition procedure which lasted until the present Act came into force. By this procedure an opposition was started by the opponent filing in the

Patent Office a Notice of Opposition summarising his case. This had to be based on the grounds which the law of the day allowed. He and the applicant then had the opportunity, in turn, to file evidence in support of their cases (as statutory declarations or affidavits), after which there was a hearing in the Patent Office court. The comptroller's decision was normally given later in writing. An appeal against the decision was possible.

The grounds of opposition allowed by the 1883 Act were 'obtaining', 'prior patenting' and 'interference' (ie conflict with a pending, and hence unpublished, patent application). The 1888 Act abolished the latter ground and introduced the ground of 'disconformity' for the special case where the opponent's invention had, as it chanced, been added to a provisional specification when the applicant had prepared his complete specification.

The 1907 Act restricted 'prior patenting' to specifications dated not more than 50 years previously (to make this ground consistent with the novelty search requirement) and added the new ground of 'insufficient and unfair description'. The 1919 Act split the ground of 'prior patenting' into 'prior publication in a British patent specification' (not more than 50 years old) and 'prior claiming', covering patent specifications of earlier priority date which had not been published at the date of the opposed application. The Act added two further grounds of opposition: 'prior publication' in any other document and 'disconformity' for the case where a convention applicant had, by chance, added the opponent's invention to the matter of his original foreign specification when preparing his British complete specification.

The 1932 Act excluded foreign patent specifications (and their abridgements) over 50 years old as prior publications, and modified the two 'disconformity' grounds by adding the possibility of opposing because an invention introduced into the complete specification had been published in the United Kingdom between the date of the basic provisional or foreign application and the date of filing the complete specification.

The grounds of opposition were further modified by the 1949 Act which retained from previous acts the grounds of 'obtaining', 'prior publication', 'prior claiming' and 'insufficient and unfair description', abolished 'disconformity', and introduced the four new grounds of 'prior use', 'obviousness', 'invention not a manner of manufacture' and 'convention application made out of time'.

The period within which a Notice of Opposition could be filed varied from the 1883 Act on as shown in the table below. This table also shows the normal times allowed for filing evidence. Since the 1932 Act, a patent applicant who wished to contest an opposition had to file a 'counterstatement' setting out his case, and the time allowed for doing this is also included in the table.

Table 11 Periods allowed for stages of opposition proceedings

| | w = week | | | | m = month | |
Year	Normal	Possible extension	Counter- statement	Opponent's evidence	Applicant's evidence	Evidence in reply
1883	2 m (a)	–	–	2 w	2 w (c)	1 w
1908	2 m (a)	–	–	2 w	2 w (c)	2 w
1932	2 m (a)	1 m	2 w	2 w	2 w	2 w
1949	3 m (b)	–	6 w	6 w	6 w	6 w
1958	3 m (b)	–	3 m	3 m	3 m	3 m

(a) From date of acceptance (b) From date of publication
(c) If no evidence filed by opponent, 3 months from date of acceptance.

The fees payable for opposing and, by each party, for attending the hearing were as shown below.

Table 12 Opposition fees

Year	Ref.	Normal opposition	Belated opposition*	Hearing
		£	£	£
1883	35	0.50	–	1.00
1907	56	0.50	2.00	1.00
1920	66	1.00	2.00	2.00
1955	93	2.00	3.00	2.00
1964	111	2.50	3.50	2.50
1969	119	2.50	3.50	3.00
1971	126	5.00	5.00	3.00
1975	134	10.00	10.00	6.00
1977	139	–	10.00	6.00

*See next section.

Although the 1977 Act does not allow pre-grant opposition proceedings, it is still possible for third parties to influence the prosecution of a patent application. Disputes about the ownership of an invention can be referred to the Comptroller, either before or after the application has been filed, in 'entitlement proceedings'. Also, when the application has been published, observations on the patentability of the invention can be sent to the Comptroller and the Examiner is bound to consider these, as well as the documents found in the Patent Office novelty search, when carrying out the substantive examination.

REVOCATION IN THE PATENT OFFICE

The 1907 Act made it possible to apply for the revocation of a patent in the Patent Office. Because the procedure was very similar to that for opposing the grant of a patent, the action was commonly called a 'belated opposition'. Originally a 'belated opposition' had to be filed within two years of the application date of the opposed patent. However, not all patent applications were accepted by then, and so 'belated opposition' of some patents was prevented. Consequently the 1932 Act changed the 'belated opposition' period to 12 months from the sealing date.

The changes in the fee for filing a 'belated opposition' are given above in the table of opposition fees. The periods for filing evidence and the hearing fees were as for ordinary oppositions. The advantage of the 'belated opposition' procedure was that it enabled someone who did not file an ordinary opposition to attack a patent much more cheaply than by petitioning the court for its revocation.

It is still possible, under the 1977 Act, to apply for the revocation of a patent at the Patent Office. The procedure is very similar to the 1949 Act procedure in 'belated oppositions' but it is no longer essential to apply within one year from grant, and the grounds on which the validity of the patent can be attacked are no longer special grounds of opposition but are the ordinary grounds of revocation.

APPEALS FROM THE COMPTROLLER

When the 1883 Act replaced the Commissioners of Patents by a Comptroller-General, the examination of patent applicants and the conduct of opposition proceedings were taken over from the Law Officers by Patent Office staff. The Law Officers were given the new task of hearing appeals from decisions of the Comptroller. No further appeal was possible.

The 1907 Act introduced some Patent Office proceedings, notably 'belated oppositions' and new 'abuse of monopoly' provisions, which could lead to the revocation of patents, and the decisions in these were made subject to appeal to a High Court Judge, not the Law Officers. Further appeal was possible in some instances.

In their report of 1931, the Sargent Committee recommended that a High Court judge should relieve the Law Officers of their work in hearing appeals from the Comptroller. They made this recommendation because of the delays caused by other duties of the Law Officers and of the possible inexperience of the Law Officers in questions relating to patents, and their possible lack of technical knowledge. The recommendation was embodied in the 1932 Act which established a Patents Appeal Tribunal consisting of a High Court judge specially appointed by the Lord Chancellor.

The powers of the Tribunal were at first similar to those of the Law Officers, and so it considered appeals from decisions concerning patent applications, not granted patents. Later, by the 1949 Act, it was empowered to hear appeals from all decisions of the Comptroller. However, in the cases where, under the 1932 Act, appeal had been possible to the High Court, the patentee was given the right of further appeal to the Court of Appeal. Also if, in opposition proceedings, the Comptroller refused to grant a patent on the ground that the invention had been prior used or was obvious, the applicant could, with leave, appeal to the Court of Appeal.

The Patents Appeal Tribunal was not a part of the High Court but was a tribunal of inferior status. Whilst this had the advantage of minimising the cost of the proceedings, it had the disadvantage (until 1971) that appeals from its decisions were only readily possible in the instances mentioned in the last paragraph. In other instances, the only possibility was to obtain leave to apply for, and then apply for, an order of *certiorari* from the Court of Queen's Bench. The order could be obtained if it could be shown that there was an error in law on the face of the record of proceedings. This method of appeal was cumbersome, costly and uncertain, and resulted in the case being reconsidered by a bench of judges less experienced in patent matters than the judge who made the appealed decision. In 1971 the need for *certiorari* proceedings was removed by the Courts Act which gave a right of appeal to the Court of Appeal from any decision of the Appeal Tribunal on the ground that it was wrong in law or in excess of jurisdiction. Leave to appeal had to be obtained from the tribunal or the Court of Appeal.

Under the present Act, all appeals from decisions of the Comptroller are made to the newly formed Patents Court, which is part of the Chancery Division of the High Court. However, the right of audience of solicitors or patent agents, which dates back to the days of appeals to the Law Officers, has been preserved so that the appeals may still be less costly than other High Court proceedings.

THE RIGHTS OF A PATENT PROPRIETOR

A valid patent gives the proprietor the right to stop others using the patented invention, a right of the kind termed legally an 'exclusive right'. The following excerpts from the wording used for letters patent under the 1852 Act give an idea of the impressive manner in which this right was expressed in the days when patents were royal grants.

'Victoria, by the grace of God of the United Kingdom of Great Britain and Ireland, Queen, defender of the faith:

Whereas A.B. hath by his petition humbly represented unto us that he is in possession of an invention for . . . , which the petitioner believes will be of great public utility; that he is the first and true inventor thereof, and that the same is not in use by any other person or persons, . . . And we, being willing to give encouragement to all arts and inventions which may be for the public good, are graciously pleased to condescend to the petitioner's request: Know ye, therefore, that we, of our especial grace, certain knowledge, and mere motion, have given and granted, and by these presents, for us, our heirs and successors, do give and grant unto the said A.B., his executors, administrators and assigns, our especial licence, full power, sole privilege, and authority, that he the said A.B. . . . his executors, administrators and assigns, and every of them, by himself and themselves, or by his and their deputy or deputies, servants or agents, or such others as he the said A.B. . . . shall at any time agree, and no others, from time to time and at all times hereafter during the term of years' (fourteen) 'herein expressed shall and lawfully may make, use, exercise, and vend his said invention within our United Kingdom of Great Britain and Ireland, the Channel Islands, and Isle of Man, in such manner as to him said A.B., . . . or any of them, shall in his or their discretion seem meet; and that he the said A.B. . . . shall and lawfully may have and enjoy the whole profit, benefit, commodity, and advantages from time to time coming, growing, accruing, and arising by reason of the said invention . . . and to the end that he, the said A.B. . . . may have and enjoy the full benefit and the sole use and exercise of the said invention . . we do require and command all and every person and persons . . . that neither they nor any of them, at any time during the continuance of the said term of fourteen years hereby granted, either directly or indirectly do make, use, or put into practice the said invention, or any part of the same . . . nor in anywise counterfeit, imitate, or resemble the same, nor shall make or cause to be made any addition thereto or subtraction from the same, whereby to pretend himself or themselves the inventor or inventors, devisor or devisors thereof, without the consent, licence, or agreement of the said A.B. . . . in writing under his or their hands and seals first had and obtained in that behalf, upon such pains and penalties as can or may be justly inflicted on such offenders for their contempt of this our royal command, and further to be answerable to the said A.B. . . . according to law for his and their damages thereby occasioned . . '

If someone carries out an act which a patent proprietor considers is covered by the patent, he can sue that person in a court of law for infringement of his rights. The court will determine whether the act complained of really is an infringement, whether the patent is valid (at least in part) and, if it is, what relief the proprietor is to be awarded.

The types of act which may constitute infringements of patent rights were settled under the former law by many decisions and were not altogether consistent with the wording of the letters patent. The new law, as laid down by the 1977 Act, defines infringement as follows: '. . a person infringes a patent for an invention if, but only if, while the patent is in force, he does any of the following things in the United Kingdom in relation to the invention without the consent of the proprietor of the patent, that is to say —

(a) where the invention is a product, he makes, disposes of, offers to dispose of, uses or imports the product or keeps it whether for disposal or otherwise;

(b) where the invention is a process, he uses the process or he offers it for use in the United Kingdom when he knows, or it is obvious to a reasonable person in the circumstances, that its use there without the consent of the proprietor would be an infringement of the patent;

(c) where the invention is a process, he disposes of, offers to dispose of, uses or imports any product obtained directly by means of that process or keeps any such product whether for disposal or otherwise.' A person also infringes a patent 'if, while the patent is in force and without the consent of the proprietor, he supplies or offers to supply in the United Kingdom a person other than a licensee or other person entitled to work the invention with any of the means, relating to an essential element of the invention, for putting the invention into effect when he knows, or it is obvious to a reasonable person in the circumstances, that those means are suitable for putting, and are intended to put, the invention into effect in the United Kingdom.'

Under the 1949 and earlier Acts it would not have been an infringement merely to keep a patented product, or to supply means for putting an invention into effect, this last activity being known as 'contributory infringement'.

A number of acts are listed in the new Patents Act as not constituting infringements of a patent. These include an act done privately for purposes which are not commercial; an act done for experimental purposes relating to the subject matter of the invention; the preparation in a pharmacy of a medicine prescribed for an individual; and the use of the invention in connection with a foreign ship, aircraft, hovercraft or vehicle which temporarily or accidentally enters the realm.

In its early years, the patent system was under the control of the Privy Council. Because patents were granted under the royal prerogative, disputes concerning them might be considered by the court which administered that prerogative, the Court of Star Chamber. This consisted of members of the King's Council, the Chancellor, Treasurer and Privy Seal, and common law judges. Because this court used inquisitorial procedures, it became much feared and disliked, and its abolition in 1641 was one of the first legislative acts of the Long Parliament.

Queen Elizabeth I's proclamation of 1601, and later the Statute of Monopolies, gave anyone aggrieved by a monopoly the right to take his case to one of the common law courts. Of these there were three. The oldest was the Court of Exchequer established in the early 12th century to deal with fiscal matters. The judges of this court were called Barons (originally spelled Barones), the principal judge being the Chief Baron. The two other common law courts were established following Magna Carta of 1215 in which King John promised that 'common pleas shall not follow our court but shall be held in a certain place'. The one, the Court of Common Pleas, was formed from the Coram Justiciares de Banco (the Court of the Justices of the Bench); the other, the Court of King's Bench, was formed somewhat later from the Coram Rege (the King's Court) and exercised surveillance over crimes and other pleas of the Crown and could hear appeals from the Court of Common Pleas. Each of these courts had a Chief Justice and a number of puisne (pronounced 'puny') judges.

The common law courts were unable to consider all kinds of case, or to offer all kinds of legal remedy, and so by the early fourteenth century people unable to obtain justice at common law were taking their cases to the King in Council, petitioning him to exercise his prerogative in their favour. Originally he heard these cases himself, but as they became more numerous he referred them to his principal minister, the Chancellor, who was secretary of the Council. From the fifteenth century onwards, cases were increasingly addressed directly to the Chancellor, in the Chancery, and there grew an independent jurisdiction, called Equity, whereby the Council could see that the law was enforced fairly and effectively. In 1474 there is the first recorded instance of the Chancellor issuing a decree in his own name, rather than in the name of the King, and so this is the date at which the Court of Chancery, a court independent of the King and of the Council, may be considered to have been created.

The chief assistants of the Chancellor came to be known as 'Masters' in the fourteenth century, and the foremost of these, the Master of the Rolls, effectively became a second judge of the Court of Chancery in 1729.

In the later thirteenth century, two distinct types of assembly, one feudal and the other representative and national, were brought together, the union ultimately producing the House of Lords and the House of Commons. In the reign of Edward I (1272-1307) the records of the 'Parliament Rolls' started to include hearings of appeals from the Court of King's Bench, and this was the beginning of the process which led to the High Court of Parliament, and subsequently the House of Lords, becoming the highest court of common law. After the Restoration of King Charles II in 1660, the Lords also assumed ultimate jurisdiction in Equity.

Whilst the Court of Common Pleas acquiesced to the superiority of the Court of King's Bench, the Court of Exchequer did not and so a new court, the Court of Exchequer Chamber, was established in 1357 to hear appeals from the Exchequer. Originally, appeals from decisions of the Court of King's Bench could only be made to the High Court of Parliament, but in 1585 a court, also called the Court of Exchequer Chamber, was set up which could review most of its decisions. There was a third Court of Exchequer Chamber formed of judges of all the courts who met to consider difficult points of law. Errors in law of the Court of Chancery were taken to the House of Lords.

These arrangements continued until 1830 when yet another Court of Exchequer Chamber was founded, the earlier courts of that name being abolished, which could hear appeals from all the common law courts.

The courts were reorganised by the Judicature Acts of 1873-5. These established a Supreme Court of Judicature comprising the High Court of Justice and the Court of Appeal. The High Court originally had five divisions: Queen's Bench, Common Pleas, Exchequer, Chancery, and Probate, Divorce and Admiralty, but in 1880 the Common Pleas and Exchequer Divisions were abolished by being merged with the Court of Queen's Bench. In 1876, the Appellate Jurisdiction Act provided for the creation of salaried life peers to hear appeals in the House of Lords. At such a hearing there have to be present at least three of the following: the Lord Chancellor, the Lords of Appeal in Ordinary (generally appointed from the ranks of the Court of Appeal), and such peers who hold, or have held, high judicial office.

After the reorganisation of the courts; patent cases were usually heard in the Chancery Division. The 1977 Act established a special Patents Court, as part of the Chancery Division, where patent actions (including appeals from decisions of the Comptroller) now take place.

Appeals from the High Court normally go to the Court of Appeal, Appeals from there go to the House of Lords, which has ultimate jurisdiction. The Administration of Justice Act of 1969 made it possible, with leave, for an appeal considered to involve an important point of law, to be made directly from the High Court to the House of Lords to avoid the additional delay and expense of proceedings in the Court of Appeal.

THE INFRINGEMENT ACTION

Despite Queen Elizabeth's and subsequent proclamations, and the Statute of Monopolies, almost all patent litigation continued to be conducted before the Privy Council. It was not until a dispute of 1753 between Lord Mansfield and the Privy Council over a petition to revoke Dr James's 1747 patent for a fever powder that the Council compelled patentees to take the common law remedy if their patents were infringed.

Patent litigation before the 1852 Act was much complicated by the facts that only the Court of Chancery could grant an injunction (to stop an infringer) or cancel a patent, and only one of the common law courts could decide on whether a patent was infringed and whether it was valid. It was therefore usual in an infringement action, first to attempt to obtain an interim ('interlocutory') injunction to restrain the defendant from continuing his alleged infringement; then to have the questions of infringement and validity tried in a common law court, the process possibly involving a retrial or one or more appeals; and finally if one had succeeded in the infringement action, to return to the Chancery to obtain a permanent injunction and possibly have the patent cancelled. These procedures were complex, slow and expensive.

The trial in the court of common law (usually the Court of King's Bench or Common Pleas, but sometimes the Court of Exchequer) was normally before a single judge with a jury, the decision being reviewed by all the judges of the court (*in banco*). This procedure for deciding issues before a single judge with a jury had been introduced by the Statute of Nisi Prius in 1285 to avoid the problem of transporting local juries to Westminster. The sheriff of a county was directed to secure the attendance of a jury at Westminster, to conclude a case started in the Court of King's Bench or Common Pleas, 'unless before' (*nisi prius*) that day the justices of assize should visit the county. The issues of fact were decided by the local jury before the justices, and their verdict was added to the record of the case in Westminster where judgement was delivered.

In a patent case tried before a jury, the jury decided on the circumstances attending the alleged infringement, the true meaning of technical terms used in the specification and the identity or otherwise of the inventions concerned. The court construed the specification and instructed the jury on the relevant points of law. The jury then gave their verdict.

In the seventeenth century it became possible for an unsuccessful defendant to move for a new trial in the same court, but the necessary permission (rule *nisi*) was not always granted. An alternative possibility, only open where a mistake in law appeared in the official record of the case, was for the defendant to appeal by means of a writ of error to the appropriate appeal court.

The procedural complexity of patent litigation was reduced in the mid-nineteenth century by the Patent Law Amendment Act, 1852, which allowed a court of common law to make, during an infringement action, any order for an injunction, inspection or account which it thought fit, and by the Evidence Act, 1851 which allowed such a court to compel the disclosure of evidence exclusively within the knowledge of the defendant. It thus became possible for patent cases to be both commenced and determined in the superior courts of common law. The Common Law Procedure Act of 1854 allowed a judge, with the consent of the parties, to try issues of fact without a jury. The use of a jury in patent cases declined rapidly and after the 1883 Patents Act, juries were only used in actions concerning threats or fraudulent behaviour.

Despite the changes in procedure, the fundamental duties of the court in a patent infringement action have remained the same: to decide if infringement has occurred and, if it has, to decide whether the patent is valid (at least in part) and so determine whether the patent proprietor is entitled to any relief. In deciding on infringement, the court first construes the specification to decide exactly what it means. Evidence about the significance of any technical terms and the sufficiency of the decription is provided by experts in the field. They often arrange for experiments to be carried out to resolve technical questions which arise. On the basis of the evidence and the language of the specification the court then decides, if it can, what the scope of the protection afforded by the patent is. Originally this was done by considering the specification as a whole, any claims present being used more to indicate which portions of his specification the patentee considered described his invention than to define with precision the boundary of his monopoly. However, as the drafting of claims developed, the courts took progressively more notice of them, and since the early part of the present century patentees have often been bound by the wording of their claims.

Yet there have always been cases where a defendant has clearly indulged in profitable ventures based on a patented invention without infringing any claim of the patent, usually by employing a substitute for some feature incorrectly considered, by the patentee, to be essential. In such a case the court might rule that the defendant had taken the 'pith and marrow' of the invention and had thereby infringed. As recently as 1973, an activity not covered by the claims of a patent was held to be an infringement, the judgement being upheld by the House of Lords. The defendant had imported a derivative ('Hetacillin') of a patented drug ('Ampicillin') into this country. Because the derivative immediately yields the drug in the body, the court held that the defendant had deprived the patentee of part of the benefit of his invention and had thereby infringed the patents concerned, even though the derivative was not covered by any claim.

As recommended by the Banks committee, the 1977 Act recognises that infringements may not be covered by the claims of an infringed patent. It does so by reference to the Protocol on the interpretation of the relevant article (No. 69) of the European Patent Convention which provides that the extent of the protection conferred by a European patent is not as narrow as 'that defined by the strict, literal meaning of the wording used in the claims, the description and drawings being employed only for the purpose of resolving an ambiguity found in the claims', or as wide as a person skilled in the art who considered the description and drawings would deduce to have been contemplated by the patentee, but is between these extremes 'combining a fair protection for the patentee with a reasonable degree of certainty for third parties'.

The defendant in an infringement action has normally always challenged the validity of the patent which he has allegedly infringed. Until the 1907 Act, however, a successful defence based on invalidity only resulted in revocation of the patent if the defendant separately petitioned for revocation. The 1907 Act allowed a defendant in an infringement action to counterclaim for revocation of the patent, or patents, concerned in accordance with Rules of the Supreme Court, Until the 1932 Act a defendant had to obtain the permission ('fiat') of the Attorney-General to counterclaim in this way.

AMENDMENT AND PARTIAL VALIDITY OF PATENTS

In the early years of the patent system, a patent could not be amended (except for verbal or clerical errors). Also, if it was defective in any significant way, it was invalid and so useless. When the validity of a patent for an obviously meritorious invention was tried, the court would sometimes uphold the patent by disregarding minor errors in the specification. Nevertheless, the patents for some important inventions were invalidated for what would later have been considered insubstantial reasons. For instance, patents of 1801 and 1803 for the Fourdrinier type of paper-making machine were revoked because the title of a specification describing subsequent improvements, which had been filed when the terms of the patents had been extended by Act of Parliament, was not consistent with the machine described in the specification of the earlier patent.

It was only by the 1835 Act that a patentee (and by the 1839 Act an assignee) was allowed to amend his patent. He did so by filing with the Clerk of Patents (of England, Scotland or Ireland), after obtaining leave of one of the Law Officers of the country concerned, a disclaimer or a memorandum of alteration. The application to amend could be opposed by any person entering a caveat and the dispute was settled by the Law Officer.

74

A patent proprietor is most likely to wish to amend his patent in two instances: before starting an infringement action if he is aware of something, usually a prior publication, likely to render the patent invalid; and during an infringement or revocation action if the court finds that the patent is invalid.

For amending in the first of these instances, the 1835 Act procedure was satisfactory, and was retained when the patent law was reformed in 1852. The 1883 Act revised the procedure making it very like that which still applies. A request to amend a patent is made to the Comptroller at the Patent Office. The request and details of the proposed amendments are advertised to give interested parties the chance to oppose. The allowability of the amendments is then decided by the Comptroller, there being a hearing if an opposition has been entered. It is possible to appeal against the Comptroller's decision.

Before the 1977 Act, amendments could be by way of disclaimer, correction or explanation; under the 1977 Act, only amendments by way of disclaimer are clearly allowable. The practice concerning other types of amendment has yet to be established.

For amending a patent which was, or was about to be, the subject of an infringement action, the 1835 Act procedure was not helpful to patentees because once an action was pending, any amendments filed were not receivable in evidence in the action. The 1852 Act improved matters somewhat because the Law Officer could give permission for an action to be brought for infringements of an amended patent which had been committed before the amendments had been filed.

The situation was much improved by the 1883 Act which allowed the court, in an infringement action, both to order a patentee to amend his patent, by disclaimer, in the Patent Office, and to award him damages for infringements of the unamended patent if satisfied that 'his original specification was framed in good faith and with reasonable skill and knowledge'. The 1919 Act allowed the court to consider the validity of a specification claim by claim so that relief could be granted for infringement of valid claims without regard to the invalidity of other claims. As previously, the court considered whether the patentee had behaved in good faith when deciding on costs, the date from which damages should be reckoned and the amendments which the patentee should be allowed to make so as to render his patent valid.

The 1919 Act provisions made it very desirable for a patentee to have a series of claims of progressively narrowing scope because in this way he had the best chance of having at least one claim narrow enough to avoid any prior publications first discovered in the infringement proceedings and yet wide enough to cover the alleged infringement.

Under the present law, a patent can be partially valid (the 1977 Act is the first Act to use this term) without there being the restriction that the valid part of the patent has to be the subject of a single claim. The need for long series of claims seems therefore to have been lessened.

THE REMEDIES FOR INFRINGEMENT

There are five main remedies available to a patent proprietor who is successful in an infringement action.

(1) An injunction. This is a court order restraining someone from continuing to do something: in the present case restraining the infringer from continuing to infringe the patent. In many instances he will have been stopped before the trial, the patent proprietor having made out a good enough *prima facie* case for the validity of his patent to obtain an interim ('interlocutory') injunction. Until the 1852 Act, injunctions could only be granted by the Court of Chancery.

(2) Damages. These constitute monetary compensation intended to restore the patent proprietor to the position he would have been in if the infringement had not taken place. Until the 1852 Act, damages could only be awarded by a Court of Common Law.

(3) An account of profits. This is an alternative to damages and, as its name indicates, is an account of the profits made by the infringer as the result of his infringement. An account may be more difficult to assess than damages, but is likely to be preferable for a proprietor who has not himself attempted to exploit his patent. Until the 1852 Act, an account could only be ordered by the Court of Chancery. The 1919 Act took away the possibility of obtaining an account, but it was restored by the 1949 Act because the Swan Committee considered that it was undesirable that a patent proprietor might be compelled, if damages were assessed, to disclose to the infringer commercial information that he would rather have kept to himself.

(4) A certificate of contested validity. To reduce the chance of a patent proprietor having to defend the validity of his patent more than once, he has been able since the 1835 Act to obtain a certificate of contested validity if his patent has been found valid in a first action. The certificate would entitle him, if successful in a later action, to more realistic costs, the 1835 Act specifying triple costs, and the 1852 and later Acts specifying actual costs between attorney or solicitor and client.

(5) Delivery up or destruction of infringing articles. A defendant found guilty of infringing a patent is restrained by the court from selling any infringing articles both before and after the patent has expired. He can be ordered to deliver up his entire stock of the articles so that they can be rendered unsaleable by marking or destruction. In some cases they can be modified so as no longer to infringe. The ownership of the articles is not transferred to the patent proprietor so he cannot sell them himself.

INNOCENT INFRINGEMENT AND THE MARKING OF PATENTED ARTICLES

Since the 1907 Act, a person who has been found to have made articles covered by a patent belonging to someone else, but who can prove that he had no reason to know of the existence of that patent, has not been liable to pay damages for infringement. The Act specified that the presence on articles made by the patentee of the word 'patent, or 'patented' without the year and number of the patent did not constitute notice of the existence of the patent, and similar provisions have been included in subsequent Patents Acts. Since 1907, therefore, it has been desirable for a patentee to mark patented articles with the appropiate application or patent number. Marking should be discontinued when the patent expires because since the 1835 Act anyone who has falsely represented anything sold by him as a patented product has been liable to a fine.

CONFIRMATION OF LETTERS PATENT

The 1835 Act allowed a patentee who was found in any suit or action not to have been first inventor of the patented invention to petition the King in Council for the patent to be confirmed. The petition was advertised and could be opposed. The Privy Council would only report in favour of the petition if, when the patent was granted, the patentee had believed himself to be the first inventor and the invention was not in use, and had never been generally used. There seems to have been only one successful petition for confirmation: Baron Heurteloup's 1834 patent for improvements in fire-arms was confirmed after he had discovered in the British Museum copies of a book describing a similar invention patented in France.

These curious provisions were retained by the 1852 Act, but not by that of 1883.

REVOCATION BY THE COURT

If a person wishes to practise an invention covered by a patent which he believes is invalid, he can only ignore the patent at the risk of having his activities stopped by legal proceedings. And yet he may well not wish to apply for a licence under the patent because in doing so he is tacitly admitting the possibility of the patent being at least partly valid. In these circumstances he may decide to apply to the court for the patent to be revoked.

Before the Patents Act of 1883, an action to repeal a patent was by means of a writ of *scire facias*. This issued out of the Petty Bag Office of the Court of Chancery and directed the sheriff of the county in which the patentee dwelt to give notice (*scire facias*) to the patentee that the Sovereign had been informed that his patent was invalid and requiring him to come into court to show why it should not be cancelled. The person attacking the patent (the prosecutor) had to give a bond to secure the payment of costs to the patentee if the patent were upheld. The Attorney-General, on behalf of the Queen, filed a declaration against the patentee, and the validity of the patent was then tried, usually with a jury in the Court of King's (or Queen's) Bench. The verdict was returned into the Chancery where, if the verdict was against the patentee, the patent was cancelled.

The 1883 Act replaced the *scire facias* procedure by a procedure involving petitioning the Court. Unless the person sued for infringement was either a person, possibly the true inventor, from whom the invention had been fraudulently obtained, or a person who had commercially used the invention before the date of the patent, he had to obtain the fiat (consent) of the Attorney-General before he could petition for revocation of the patent. The fiat was always granted and so the Sargent Committee, and later the Swan Committee, considered the requirement for obtaining the fiat an unnecessary source of expense and delay. The requirement was finally abolished by the 1949 Act.

THREATS

There has always been the temptation for a patent proprietor to use for his own ends the threat of infringement proceedings. For instance, he might be tempted to frighten the customers of a competitor to stop them buying a competitor's goods. The provisions of the patent law for discouraging threats have varied as will now be outlined.

Before the 1883 Act a threatened person could only obtain an injunction restraining the threats if he could show that they were made in bad faith and that he had not infringed the patent (which was presumed valid). He could only obtain damages if he could prove that the threats were a malicious attempt by the patentee to injure him.

The 1883 Act improved matters, making it possible for a threatened person to obtain an injunction and damages even if the threat had been made in good faith, but the action could be avoided if the person making the threat started and prosecuted an infringement action with diligence.

The 1932 Act allowed a threats action to be brought against anyone, not — as previously — only against someone claiming to be the patentee, and removed the possibility of avoiding the action by pursuing infringement proceedings. The 1949 Act made little change in the law but rendered its provisions less effective by declaring that a mere notification of the existence of a patent did not constitute a threat of proceedings. The result of this declaration was that patentees only dared send mere notifications which often proved not to be clear enough warnings of infringement.

The Banks Committee were convinced that the threats provisions were inadequate, and the changes which they recommended were in substance embodied in the 1977 Act which provides that 'proceedings may not be brought . . . for a threat to bring proceedings for an infringement alleged to consist of making or importing a product for disposal or of using a process'.

ABUSE OF MONOPOLY AND COMPULSORY LICENCES

Until the mid eighteenth century, patents could be invalidated if the inventions concerned were not put into practice within a certain period. Also, letters patent contained, until their form was revised by the 1908 Rules, the proviso that they would be void if it appeared to at least six members of the Privy council that, amongst other things, they were prejudical or inconvenient to His, or Her, Majesty's subjects. Even after the time when publication, rather than working, of an invention became the main reason for granting a patent, there remained the fear that a patentee might use his patent merely to stop others working the invention and have no intention of exploiting it himself. The 1883 Act therefore empowered the Board of Trade to order the grant of a compulsory licence under a patent on any of the grounds (1) that the patented invention was not being worked, (2) that the needs of the public were not being met, and (3) that the person seeking the licence was being prevented by the patent form working another patented invention.

The 1902 Act transferred the power to grant a compulsory licence to the Judicial Committee of the Privy Council and gave them power to revoke a patent three or more years old if they considered granting a compulsory licence under it was not an adequate remedy. The 1907 Act transferred these powers to the court and additionally gave the Comptroller the power to revoke a patent any time after four years from its date if the invention was being exercised mainly or wholly abroad.

The 1919 Act introduced the present arrangement by which 'abuse of monopoly' matters are dealt with by the Comptroller. It gave him the additional remedy of ordering a patent to be indorsed 'licences of right'. The 1928 Act made it allowable to apply for a compulsory licence, or other remedy, only in the case of a patent at least three years old.

Whilst successive Patents Acts have elaborated the abuse of monopoly provisions, the underlying principles have not significantly altered. The principal change introduced by the new Patents Act was the removal of the possibility of a patent being revoked if compulsory licensing failed to achieve its purpose. The fact that the provisions have been little used indicates that they have been an effective deterrent.

The grant of compulsory licences under patents for medicines is considered later.

CROWN USE

Up till 1883, the Crown could use any patented invention without the permission of the patentee and without compensating him. In fact patentees were sometimes rewarded *ex gratia*. Since the 1883 Act, a patent has been binding against the Crown as against a subject, but the Crown has nevertheless been empowered to use a patented invention without the permission of the patent proprietor, although not without paying him any compensation.

PATENTS AND MEDICINES

Patents have been associated with medicines since the early days of the patent system. The first primarily medical English patent was that granted to Nehemiah Grew in 1698 for a cheap method of making Epsom salts (magnesium sulphate), the active constituent in water from the well discovered at Epsom in 1618.

An Act of 1783, in the reign of King George III, imposed stamp duties on all proprietary medicines, whether the subject of letters patent or not, and eventually the term 'patent medicine' came to be associated with any medicine on which a stamp duty was levied. Initially, the duty was 3d for a medicine of value less than 2s 6d, 6d for one of value from 2s 6d to under 5s, and 1s for one of value 5s or more. Counterfeiting stamps was a felony punishable by death, and re-use of an old stamp carried a fine of ten pounds.

The 1783 Act was not fully effective, and was replaced by an Act of two years later which was in turn replaced by an Act of 1802.

The schedule of this Act listed numerous medical preparations on which duties were payable and, as the following few examples show, is entertaining to read.

American Alterative Pills
Birt's Martial Balsam
Chalybeate Female Pills
Doranstoff's Opodeldoc
Elixir of Longevity, or Swedish Preservative
Friend to Man
Ginger Pearls
Hewitt's Analambanic Pills
Infallible Restorative
Jebb's Elixir
Kibe Ointment
Lowther's Nervous Powders
Metallic Tractors
Nervous Cordial

Ormskirk Medicine for the Bite of a Mad Dog
Pullin's Purging Pills
Quassia Pills
Royal Tooth Powder
Samaritan Restorative
Turkey Rhubarb Lozenges
Vital Balm
Ward's Sack Emetic
Zimmerman's Stimulating Fluid.

The 1802 Act, and some subsequent legislation, lasted right until the Pharmacy and Medicines Act of 1941, which abolished stamp duties on medicines and compelled the makers of proprietary medicines to disclose the composition.

After the First World War, it was evident that the United Kingdom pharmaceutical industry lagged far behind that in Germany. To help the United Kingdom industry catch up, the 1919 Act introduced two measures designed to lessen the effectiveness of patents relating to food and medicines. The first was to prohibit the claiming of a chemical compound *per se*, ie regardless of how it had been prepared: it could only be claimed as the product of a particular process. Thus a patentee could no longer control the entire manufacture and importation of a compound, his control only extending to the compound as made by his own process. The second was to require the Comptroller to grant a compulsory licence under a food or medicine patent to anyone who seemed competent to work the invention. Thus the patentee of an invention in this field was more likely to be deprived of his right to prevent competition than someone who had patented any other kind of invention, his only comfort being the royalties he obtained from the licensed use. An application of a compulsory licence could be and often was opposed, the usual opponent being the patentee.

After the Second World War, compulsory licences were increasingly sought by companies which carried out no research and wanted to make profits by selling, at a price which might be uneconomic for the patentee, products made cheaply abroad in countries like Italy where pharmaceutical patents were not allowed. The research-based companies in this country were, not surprisingly, very much against the grant of licences for this purpose and expressed these views to the Banks Committee. In its report, the Committee recommended the abolition of the particularly harsh compulsory licensing provisions for food and medical patents, and they were in fact not included in the 1977 Act.

Another contentious topic concerning pharmaceutical patents is the way in which the Government exercises its right under the Crown use provisions to employ any patented invention it chooses. In 1961 the Minister of Health sanctioned the supply to National Health Service hospitals of quantities of five widely-used medicines obtained from unlicensed sources abroad. A United Kingdom pharmaceutical manufacturer (Pfizer) took the Ministry to court but lost by three votes to two on appeal to the House of Lords. It was thus decided that the supply of medicines to the National Health Service is use for services of the Crown. The Health Services and Public Health Act 1968 extended the scope of Crown use to cover the supply of National Health Service medicines through retail chemists.

UNAUTHORISED ASSUMPTION OF THE ROYAL ARMS

All patents granted under the law before the 1977 Act were royal grants. Evidently many patentees used to think that this fact entitled them to use the Royal Arms in connection with their goods or businesses. It did not, and the 1883 Act forbade such use, making an offender liable, on summary conviction, to a fine of up to twenty pounds. Also it has been possible since the Trade Marks Act of 1905 to obtain an injunction restraining someone from continuing unauthorised use of the Royal Arms. Corresponding provisions were included in the Patents Acts of 1907 and 1949 and the Trade Marks Act of 1938.

To be entitled to use the Royal Arms a tradesman has actually and habitually to supply goods to the Sovereign and has to obtain a warrant of appointment from the Lord Chamberlain's office.

Since the Trade Marks Act of 1875 it has not been possible to register a trade mark containing the Royal Arms (or anything closely similar).

GOVERNMENT AWARDS TO INVENTORS

Although patents have been the normal means of reward for inventors, rewards have occasionally been made in other ways. For example, a reward for a means of discovering longitude at sea was offered by an Act of 1713 and Joanna Stephens was rewarded by an Act of 1738 for a cure for stone.

A Royal Commission on Awards to Inventors of 1919 under the chairmanship, initially, of Sir Charles Sargent settled awards for both patented and unpatented inventions that had been used during the First World War. A similar commission, appointed in 1946 under the chairmanship of Lord Cohen, dealt with claims arising from the Second World War. Amongst the awards made for famous inventions were £100,000 to Sir Frank Whittle for his patents on aircraft jet propulsion; £12,000 to Sir Donald Bailey for his bridge design; and £94,600 to the team of Sir Robert Watson-Watt for their radar inventions.

82

PATENTING OF INVENTIONS SHOWN AT EXHIBITIONS

An inventor intending to exhibit an invention at the Great Exhibition of 1851 was enabled by an Act of that year to register his invention, by filing a description of it, and so obtain a period of a year in which to file a petition for a patent. Such registration was thus rather similar to the procedure introduced the following year for filing a provisional specification. An Act of 1865 allowed filing after showing an invention at a duly certified industrial exhibition and like provision for international exhibitions was made by an Act of 1870. These provisions were modified by the 1883 Act, extended by an Act of 1886 to cover exhibitions overseas and limited by the 1977 Act to cover only international exhibitions. Under the 1870 and later Acts filing had to be within six months of the exhibition opening.

PATENT OFFICE PREMISES

From 1617, an office called 'The Patent Office' had existed in Chancery Lane, London. This was the office of the Clerk of the Patents, an officer appointed by the Crown to deal with letters patent of all descriptions. For many years before the 1852 Act, two other 'Patent Offices' had existed. One, sometimes known as 'The Patent Bill Office', was under the Attorney-General and was the place, in Lincoln's Inn, London, where parchment copies of patent documents were prepared. The other, sometimes known as 'The Great Seal Patent Office', was under the Lord Chancellor and was the office in Quality Court, off Chancery Lane, where the final stages of issuing a patent were carried out. One of the earliest decisions of the Commissioners of Patents appointed by the 1852 Act was to provide a single office in which all the stages of obtaining a patent of invention could be carried out.

The new Patent Office was officially known as 'The Office of the Commissioners of Patents for Inventions' but was often referred to by the name of 'The Great Seal Patent Office' which it had absorbed. When the 1883 Act replaced the Commissioners by a Comptroller-General, the office became known officially as 'The Patent Office'.

The Patent Office has not moved since it was established late in 1852. It stands on ground which once belonged to the English Province of the Knights Templars to whom it had been made over by King Henry I in 1128. They obtained a larger site off Fleet Street in 1161 and sold their Holborn property of following year to the Bishop of Lincoln who built a mansion. This was known as Lincoln House until 1549 when it passed to the first Earl of Southampton and was renamed Southampton House. The house was demolished in 1652, and the site was developed into a residential area including a new street, Southampton Buildings. In this street, at Nos 25 and 26, offices for the Masters in Ordinary in

Chancery and for the Secretaries of Bankrupts and Lunatics were built in about 1793. The Masters were abolished by an Act of 1852, as part of Chancery reform, and it was in their ground floor chambers that the new Patent Office was set up. A dark corridor led first to the Patent Office, then to the Bankruptcy Office and finally to the Lunacy Office: as Harding comments in his book on the 'Patent Office Centenary'– 'a strangely significant juxtaposition'.

The initial Patent Office premises soon became inadequate. In 1886, an office which had been built in Jacobean style during the years 1842-3 for the Taxing Master in Chancery was, with other buildings, bought by the Comptroller. This style was adopted when the Patent Office was rebuilt and enlarged in stages over the years 1893 to 1912.

Since 1852, various proposals have been made to provide adequate accommodation for the Patent Office. Sites were considered in Chancery Lane (1857), Cursitor Street, Burlington Gardens and between Horse Guards Avenue and the Embankment (1862-75). More recently, it was proposed to build a science centre, including a new Patent Office, on the South Bank by Waterloo bridge, but this plan was abandoned and the site has been used for the National Theatre. In February 1966, the Government decided to move the Patent Office to Croydon but were persuaded to reverse this decision the following year. The Patent Office thus remains at 25 Southampton Buildings with many Examiners housed in offices within a half-mile radius.

PATENT OFFICE LIBRARY

The founding of the Patent Office Library was largely due to a suggestion made by the Prince Consort when Professor Bennet Woodcroft visited Windsor in 1853. The library was opened on March 5 1855 at Southampton Buildings, being partly housed in a dark central corridor know as the 'drain pipe'. New premises were built in 1866-7 and again from 1898-1901.

In 1966 the British Museum Library took over the administration of the library, which was renamed the 'National Reference Library of Science and Invention'. The library was again renamed, as the 'Science Reference Library' when the 'British Library' was formed in 1973. It is be moved when the new British Library building has been built in Euston Road on the site of an old railway goods depot to the west of St. Pancras station.

MODELS OF INVENTIONS AND THE PATENT MUSEUM

As with the Patent Office Library, the formation of a patent museum resulted from a suggestion made by the Prince Consort when Professor

84

Bennet Woodcroft visited Windsor in 1853. Professor Woodcroft's work in establishing the museum is best described in the light of his earlier career and so a few biographical details are now given.

Bennet Woodcroft was born in 1803, the son of a textile manufacturer. He studied chemistry under John Dalton, and later became an inventor not only, as might be expected, in the textile field but also in marine engineering. Over the years 1826 to 1853 he took out thirteen patents, the most successful of these commercially being that of 1838 for tappets for looms.

Presumably it was this patenting activity which kindled Bennet Woodcroft's great interest in the history of the patent system. He built up a collection of models of inventions and portraits of inventors and in 1838 started to compile a chronological index, a subject index, and a name index to all the patents granted since 1617. In 1843 he established himself as a consulting engineer and patent agent in Manchester, moving to London in 1846. From 1847 to 1851 he was Professor of Machinery at University College, London, and in 1851 gave evidence to the House of Lords Select Committee of Enquiry into the patent system   In the same year he published his book *Amendment of the Law and Practice of Letters Patent for Invention.*

When the Patent Law Amendment Act was passed in 1852, Professor Woodcroft was appointed Superintendent of the Specifications. His publishing work in this office has already been mentioned (see page 53). By then he had completed his indexes and their purchase by the Patent Office was authorised by the Act of 1853 (which, by the way, incorrectly spelled his first name as Bennett!). In 1854 he brought his collection of models of inventions and portraits of inventors to the Patent Office to form the nucleus of the proposed museum. The collection was moved to Kensington Palace in 1856 and thence to the South Kensington Museum (in Brompton Road, London, where the Victoria and Albert Museum now stands) when that was opened in 1857.

The control and management of the Patent Museum were transferred to the Department of Science and Art by the Patents Act of 1883 which also laid down that an inventor could be required by that Department to provide a model of his invention on payment of the cost of manufacture. Any dispute arising as to the amount to be paid was to be settled by the Board of Trade. The Department of Science and Art was merged with the Education Department in 1899 to form the Board of Education and the 1907 Patents Act provided, by a subsection which is still in force, that the control and management of the Patent Museum should remain vested in the Board of Education. The 1907 Act also confirmed, by the following subsection, that inventors could be required to provide models of their inventions. This subsection has only just been repealed, by the 1977 Act.

In 1899, Queen Victoria laid the foundation stone of new museum buildings in Brompton Road and directed that the South Kensington

Museum should henceforth be known as the Victoria and Albert Museum. When the new buildings were opened in 1909, by King Edward VII, the 'V. & A.' became a purely art museum, and the collection of scientific and technical items — including the models of inventions — was used to form a new museum, the Science Museum. This is to the west of Exhibition Road on a site which since 1864 had accommodated the Science Collection of the South Kensington Museum.

PATENT LEATHER

Patent leather is leather having a permanently glossy surface produced by a process variously referred to as enamelling, japanning, lacquering or varnishing. Originally all patent leather was black and was prepared by giving the leather several coats of a mixture of boiled linseed oil, black pigment, and a compound — such as a cobalt or manganese compound — which accelerated the drying of the oil by oxidation. Later, nitrocellulose was used in the lacquering composition, and more recently polyurethane resins.

The name 'patent leather' strongly suggests that the lacquering process was once patented but this does not seem to be the case; the name was well known by 1854, the date of the earliest relevant United Kingdom patent. The process was evidently devised in Europe in the late 18th century and worked secretly there. An American inventor, Seth Boyden (1788-1870), duplicated this and a number of other secret European industrial processes, setting up a factory in Newark, New Jersey, for the purpose in 1819. It was he who called the product 'patent leather'.

Patent leather was originally used in America for coachwork, but by the middle of the nineteenth century had become fashionable both there and in Europe for shoes.

REFERENCES

AND

BIBLIOGRAPHY

REFERENCES

Cases referred to

1567 Hastings's Patent, 1 W.P.C. 6 39

1602 Darcy v. Allin, 1 W.P.C. 1. Account in J.W. Gordon, *Monopolies by Patents;*
see bibliography 19

1776 Dollond's patent 19 April 1758, printed series No 721. 1 W.P.C. 43, Goodeve,
2nd ed 175 39

　　　Brand's patent 10 August 1771, printed series No 1771 42

1778 Liardet's patent 3 April 1773, printed series No 1040, 1 W.P.C. 53, Patent
Office bound volume of reports 42

1785 Arkwright's patent 16 December 1775, printed series No 1111. 1 W.P.C. 64,
Goodeve, 2nd ed 15................................... 35, 39

1787 Turner's patent 26 February 1781, printed series No 1281, 1 W.P.C. 77,
Goodeve, 2nd ed 470....................................... 44

1802 Tennant's patent, 23 January 1798, printed series No 2209, 1 W.P.C. 125,
1 Carpmael, 177 .... 35, 39

1810 Bainbridge's patent 2 April 1802, printed series No 2693. Goodeve, 2nd ed
30 43

1815 Wood v. Zimmer. Zink's patent 20 January 1812, printed series No. 3519.
Goodeve, 2nd ed 502....................................... 39

1817 Rex v. Metcalf. Metcalf's patent 30 September 1816, printed series No
4065. Goodeve, 2nd ed 299.................................45

1821 Brunton v Hawkes. Brunton's patent 26 March 1813, printed series No 3671.
Goodeve, 2nd ed 97....................................... 41

1823 Savory v. Price. Savory's patent 23 August 1815, printed series No 3954.
1 W.P.C. 83, Goodeve, 2nd ed 411 42

1825-7 Bloxham v. Elsee. Gamble's patents 20 April 1801 and 7 June 1803 and
Fourdrinier and Gamble's patent by 47 Geo. 3, c.131, printed series Nos
2487, 2708 and 3068. Goodeve, 2nd ed 65..................... 74

1836 Baron Heurterloup's patent 22 May 1834, printed series No 6611. Goodeve,
2nd ed 511.. 77

1835-7 Morgan v. Seaward. Galloway's patent 2 July 1829, printed series No
5805. Goodeve, 2nd ed 307.................................39

ITEMS REFERRED TO NOT IN THE BIBLIOGRAPHY

Privy Council Regulation, 1444. Proc. Privy Council. Vol. 6 pp. 318-9 15

Tomkins, A.B. *A Short Historical Review on the Law on the Protection of Industrial Property in Ireland* Trans. C.I.P.A. LXXXVII, CII (1968-9). 17

Dickens, Charles. *A Poor Man's Tale of a Patent,* from *Reprinted pieces & c.* . . . 18

Dodds, C. H. *A Room near Chancery Lane,* Household Words, Vol. XV, p.190, 1857, reprinted in CIPA Vol. 3 (No 10) 343 (1974) 18

John Kempe's grant, Patent Rolls, 5 Edw. 3, pt. 1, m.25. 18

John of Utyman's patent, 1449. Patent Rolls, 27 Hen. 6, pt. 2, m.17. Calendar, Hen. 6, Vol. 5, 1446-52, p. 255 18

Smyth's patent, 1552, Patent Rolls, 6 Edw. 6, pt. 5, m.6 18

Proclamation of 1601 by Queen Elizabeth. See Price, W.H. in bibliography. Facsimile available from Patent Office. 19

James I. Proclamation of 1610, Barker, London, Facsimile in Gordon, J.W. *Monopolies by Patents, etc.* (see bibliography) 19

James Watt, *Thoughts upon Patents, or exclusive Privileges for New Inventions* Reproduced in Robinson, E. and Musson, A.E. *James Watt and the Steam Revolution* , Adams & Dart, London, 1969. 20

Bain-Smith, T. *The First Patent Agent,* CIPA Vol. 1 (No 6) 244 (1972) 28

Newton, A.V. *On Patent Agency, its Origin and Uses,* paper read to C.I.P.A. Spottiswoode, London, 1892. 28

Robinson, E.B. *Our Charter,* Trans. C.I.P.A. LXXI C29 (1952-3) 28

Crumpe's patent, 9 January 1618, printed series No 8 29

Nasmith's patent, 3 October 1711, printed series No 387 29

Dr. James's patent, 13 November 1747, printed series No 626 72

Hulme, E.W. *Privy Council Law and Practice of Letters Patent for Invention from the Restoration to 1794.* L.Q.R. XXXIII p. 63, p. 180, (1917). 72

Nehemiah Grew's patent, 15 July 1698, printed series No 354 80

Cohen, The Rt. Hon. Lord, *Awards to Inventors* Presidential Address to The Holdsworth Club of the University of Birmingham, 1960 82

Caswell, B. *Patent Office: Historical Note*, CIPA Vol. 5 (No 3) 87 (1975) 83

Woodcroft's patent, 4 January 1838, printed series No 7529 85

Encyclopedia Americana, International Edition, Americana Corporation, New York, U.S.A. Entry for Boyden, Seth 86

OTHER SOURCES USED

Archer, P. *The Queen's Courts*, Pelican A365 2nd ed., Harmondsworth, Penguin, 1963

Burnett, J. *A History of the Cost of Living*, Pelican A 1020, Harmondsworth Penguin, 1969

Pemberton, J.E. *British Official Publications*, 2nd ed. Oxford, Pergamon, 1973

Walker, R.J. and Walker, M.G. *The English Legal System*, 3rd ed. London, Butterworths, 1972

BIBLIOGRAPHY

This bibliography is divided into five main sections, with subdivisons, as follows:

1 Statutes and rules

    (a)   Those concerned with patents for inventions
    (b)   Other statutes referred to
    (c)   Other rules referred to

2 Law Reports and digests

    (a)   Official
    (b)   Other

3 Official Publications

    (a)   On United Kingdom patent law
    (b)   On International agreements

4 Books on patent law and practice and the history of those subjects

    (a)   Text books and monographs
        (i) on former United Kingdom law and practice
        (ii) on the new United Kingdom and European patent systems
    (b)   Authors of leaflets
    (c)   Authors of works on patent law reform
    (d)   Works on international patenting

5 Selected books on inventions, their history and exploitation

The fundamental sources of information on United Kingdom patent law and practice are the statutes enacted by Parliament, the rules and orders made under these, and the many decisions made by the courts in settling cases on patent applications and patents. These sources are given in Sections 1 and 2.

The development of patent law and practice has been influenced by the reports of committees and by international agreements entered into by the United Kingdom. These reports and agreements have been published officially and are referred to in Section 3.

The books referred to in Section 4 are secondary sources of information although some include primary sources, reproducing statutes and rules current at the date of publication. Text books are valuable because they give details of practice which cannot be deduced from the statutes and rules, and because they describe the court decisions considered most suitable for illustrating and interpreting the law.

Section 5 gives references to a few recent books on inventions, a topic with which the present work is only incidentally concerned.

1   STATUTES AND RULES
(a) Those concerned with patents for inventions.

1   18 Hen. 6 c.1   (dating of letters patent)   1439

2   6 Hen. 8 c.15   (Crown grants)   1514

3   27 Hen. 8 c.11   (Clerks of the Signet and Privy Seal)   1535

4   21 Jac. 1 c.3   The Statute of Monopolies 1623

5   5 & 6 Will. 4. c.83   1835

6   Rules of Practice before the Attorney - and Solicitor-General   1835

7   Rules to be observed in Proceedings before the Judicial Committee of the Privy Council   18 Nov and 21 Dec 1835

8   2 & 3 Vict. c.67   1839

9   7 & 8 Vict.c.69   The Judicial Committee Act, 1844

10   Rule of Attorney-General   1850

11   14 & 15 Vict. c.8   The Protection of Invention Act, 1851

12   15 & 16 Vict. c.6   1852

13   15 & 16 Vict. c.83 The Patent Law Amendment Act, 1852

14   First set of Rules and Regulations 1st Oct. 1852

15   Order (concerning specification and drawings) 1st Oct. 1852

16   Order (concerning Law Officers' fees)   1st Oct. 1852

17   Second Set of Rules and Regulations   15th Oct. 1852

18   Order of Lord High Chancellor 15th Oct. 1852

19   16 & 17 Vict. c.5 (concerning stamp duties etc.)   1853

20   16 & 17 Vict. c.115   1853

21   Third Set of Rules and Regulations   12th Dec. 1853

22   Rule of Lord High Chancellor   17th July 1854

23   Rules (concerning communication applications) 23rd Feb. 1859

94

96

97	S.I. 1957, No 1657	Register of Patent Agents (Amendment) Rules, 1957
98	S.I. 1958, No 73	The Patents Rules, 1958
99	S.I. 1959, No 278	Patents Appeal Tribunal Rules, 1959
100	S.I. 1959, No 524	The Patents (Amendment) Rules, 1959
101	S.I. 1959, No 1569	Register of Patent Agents (Amendment) Rules, 1959
102	9 & 10 Eliz, 2 c.25	Patents & Designs (Renewals, Extensions and Fees) Act, 1961
103	S.I. 1961, No 1016	Patents Appeal Tribunal (Amendment) Rules, 1961
104	S.I. 1961, No 1185	The Patents (Amendment) Rules, 1961
105	S.I. 1961, No 1499	Patents (Fees Amendment) Order, 1961
106	S.I. 1961, No 1619	The Patents (Amendment No. 2) Rules, 1961
107	S.I. 1962, No 2730	The Patents (Amendment) Rules 1962
108	S.I. 1963, No 1982	The Patents (Amendment) Rules 1963
109	S.I. 1964, No 45	The Public Offices Fees (Patents, Designs and Trade Marks) Order 1964
110	S.I. 1964, No 228	The Patents (Amendment) Rules 1964.
111	S.I. 1964, No 1337	The Patents (Amendment No. 2) Rules 1964
112	S.I. 1964, No 2029	Register of Patent Agents (Amendment) Rules 1964
113	S.I. 1966, No 1482	The Patents (Amendment) Rules 1966
114	S.I. 1967, No 392	The Patents (Amendment) Rules 1967
115	S.I. 1967, No 1171	The Patents (Amendment No. 2) Rules 1967
116	S.I. 1968, No 1389	The Patents Rules 1968
117	S.I. 1968, No 1702	The Patents (Amendment) Rules 1968
118	S.I. 1968, No 1741	Register of Patent Agents (Amendment) Rules 1968
119	S.I. 1969, No 482	The Patents (Amendment) Rules 1969
120	S.I. 1969, No 500	The Patents Appeal Tribunal (Amendment) Rules 1969

121	S.I. 1969, No 1706	The Patents (Amendment No. 2) Rules 1969
122	S.I. 1970, No 529 (L13)	The Patents & Registered Designs Appeal Tribunal Fees Order 1970
123	S.I. 1970, No 955	The Patents (Amendment) Rules 1970
124	S.I. 1970, No 1074 (L22)	The Patents Appeal Tribunal (Amendment) Rules 1970
125	S.I. 1970, No 1953	The Patents (Fees Amendment) Order 1970
126	S.I. 1971, No 263	The Patents (Amendment) Rules 1971
127	S.I. 1971, No 394	The Patents Appeal Tribunal (Amendment) Rules 1971
128	S.I. 1971, No 1917	The Patents (Amendment No. 2) Rules 1971
129	S.I. 1972, No 1940(L28)	The Patents Appeal Tribunal Rules 1972
130	S.I. 1973, No 66	The Patents (Amendment) Rules 1973
131	S.I. 1973, No 164(L1)	The Patents Appeal Tribunal (Fees) Order 1973
132	S.I. 1974, No 87	The Patents (Amendment) Rules 1974
133	S.I. 1974, No 2145	The Patents (Fees Amendment) Order 1974
134	S.I. 1975 No 371	The Patents (Amendment) Rules 1975
135	S.I. 1975 No 891	The Patents (Amendment No. 2) Rules 1975
136	S.I. 1975 No 1021	The Patents (Amendment No. 3) Rules 1975
137	S.I. 1975 No 1262	The Patents (Amendment No. 4) Rules 1975
138	S.I. 1975 No 1467	The Register of Patent Agents (Amendment) Rules 1975
139	1977 c.37	Patents Act 1977
140	S.I. 1977 No 2090 (C70)	The Patents Act 1977 (Commencement No. 1) Order 1977
141	S.I. 1978 No 216	The Patents Rules 1978
142	S.I. 1978 No 586 (C14)	The Patents Act 1977 (Commencement No. 2) Order 1978.
143	S.I. 1978 No 621	The Patents Act 1977 (Isle of Man) Order 1978
144	S.I. 1978 No 1093	The Register of Patent Agents Rules 1978

(b)   Other statutes referred to,

Many other statutes have affected patent law and procedure. Those mentioned in the text are included in the list below. Others may be found from the 'Index to the Statutes' published annually in two volumes by Her Majesty's Stationery Office.

13 Ed. 1 c.30 Justices of nisi prius, etc., 1285   72

13 Ann. c.14 Discovery of longitude at sea, 1713   82

12 Geo. 2 c.23 Reward to Joanna Stephens . . . ,1738   82

15 Geo. 3 c.61 Steam Engine Act, 1775   63

23 Geo. 3 c.62 Stamp Duties on Medicines, 1783   80

33 Geo. 3 c.18 Acts of Parliament (Commencement) Act, 1793   19

42 Geo. 3 c.56 Medicines Stamp Act, 1802   80

5 & 6 Wm. 4 c.62 Statutory Declarations Act, 1835   15

11 & 12 Vict. c.94 Court of Chancery Offices Act, 1848   53

12 & 13 Vict. c.109   The Petty Bag Office and Enrolment in Chancery Amend-
                ment Act, 1849   53

14 & 15 Vict. c.99 Evidence Act, 1851   73

17 & 18 Vict. c.125 Common Law Procedure Act, 1854   73

36 & 37 Vict. c.66   Supreme Court of Judicature Act, 1873   71

37 & 38 Vict. c.83   Supreme Court of Judicature (Commencement) Act, 1874   71

38 & 39 Vict.c.77   Supreme Court of Judicature Act, 1875 71

38 & 39 Vict. c.91   Trade Marks Registration Act, 1875   82

39 & 40 Vict. c.59   Appellate Jurisdiction Act, 1876   71

40 & 41 Vict. c.41   Crown Office Act, 1877   14

56 & 57 Vict. c.66   Rules Publication Act, 1893   22

5 Ed. 7 c.15 Trade Marks Act, 1905   82

1 & 2 Geo. 6 c.22 Trade Marks Act, 1938   82

4 & 5 Geo. 6 c.42 Pharmacy and Medicines Act, 1941 81

9 & 10 Geo. 6 c.36 Statutory Instruments Act, 1946 22

1968 c.46 Health Services and Public Health Act 1968 82

1969 c.58 Administration of Justice Act 1969 71

1971 c.23 Courts Act 1971 67

c) Other rules referred to

S.I. 1970 No 1537 The Secretary of State for Trade and Industry Order 1970
22

S.I. 1974 No 692 The Secretary of State (New Departments) Order 1974 22

2 LAW REPORTS AND DIGESTS

Reports of the early patent cases were scattered amongst many law journals. To combat this inconvenience a number of authors prepared, in varying detail, collected reports. Probably the most famous of these are the ones by Thomas Webster. He had helped in preparing the consolidated bill which became the Patent Law Amendment Act of 1852.

After the 1883 Act, publication of the official 'Reports of Patent Design & Trade Mark Cases' (R.P.C.) was started, Volume 1 being for the year 1884. After 1955 (Volume 72) the volume numbers were discontinued, the volumes then being identified solely by the year of publication.

In November 1963 a new journal, 'The Fleet Street Patent Law Reports', was started to provide more up to date coverage than was possible in the official reports. For speed, it was produced by copying typescript. The journal was taken over by Thomson Publications and with the January 1975 issue, numbered Vol. 1 No. 1, appeared in more formal guise. Many of the cases which appear in the Fleet Street Reports are subsequently reported in the R.P.C.

(a) Official

Reports of Patent, Design, Trade Mark and Other Cases Vol. 1, 1884 to Vol. 72, 1972. Volumes from 1973 on are indicated by the year of publication. Each year a digest of the cases reported during the previous year is published.

Digest of Patent cases bearing upon the Practice at the Patent Office reported in Vols I to XLIX of the R.P.C. etc. The Patent Office, London, 1933 (reprinted 1947).

Digest of Patent Cases bearing upon the Practice at the Patent Office reported in Vols. L to LXIII of the R.P.C. The Patent Office, London, 1947

Consolidated Table of Cases reported in Vols. I to LXX of the R.P.C. The Patent Office, London, 1954

100

Digest of the Patent, Design, Trade Mark & Other Cases reported in Vols. I to LXXII of the R.P.C. Vol. I Cases relating to Patents; Vol. II Cases relating to Patents and Designs. The Patent Office, London, 1959

(b) Other

Carpmael, W.,	*Law Reports of Patent Cases*; Macintosh, London Vol. 1, 1843, Vol. 2 1851
Davies, J.,	*A Collection of the most important cases respecting Patents of Invention, since the Statute for restraining monopolies;* Reed, London, 1816
Goodeve, T.M.,	*Abstract of Reported Cases relating to Letters Patent for Inventions*; lst ed. 1876, 2nd ed. 1884, Sweet, London
Griffin, R.,	*Abstract of Reported Cases relating to Letters Patent for Inventions (between the years 1884 to 1886 inclusive).*; Sweet, London, 1887
	Patent Cases decided by the Comptroller-General and Law Officers of the Crown in 1887; Waterlow, London, 1888
Haseltine, G.,	*British Letters Patent (notes of cases of Opposition before the Law Officers)* London, 1875
Higgins, C.,	*A Digest of the Reported Cases relating to the Law and Practice of Letters for Inventions*; lst ed. 1875, Clowes, London Supplement, 1880 2nd ed. 1890, with Jones, G.M.E.
Macrory, E.,	*Reports of Cases relating to Letters Patent for Inventions;* London, 1855
Pirani, S.G.,	*Index of Patent Cases from 1884 – 1909 with Indexes of Subject-Matter*; Sweet and Maxwell, London 1910
Prince, A.,	*The Record of Patent Inventions with Law Reports of Patent Cases*; Vol. 1, 1842-3
Webster, T.,	*Reports and Notes of Cases on Letters Patent for Inventions*; Thomas Blenkarn, London. Vol. 1 1844, Vol. 2 1855

The Fleet Street Patent Law Reports; Fleet Street Patent Law Reports Ltd., From Feb. 1978 European Law Centre Ltd. 1973 to 1974, issues numbered annually. Annual volumes from 1975 (Vol. 1) onwards.

Many official publication are Parliamentary papers presented to Parliament *By Command of Her Majesty*, and hence known as *Command Papers*. Since 1833 these have been numbered in series, the numbers of the different series being distinguished by the prefix and, until 1922, square brackets as shown below.

Series	Paper Numbers	Years of Presentation
1	[1] − [4222]	1833 − 1869
2	[C. 1] − [C.9550]	1870 − 1899
3	[Cd. 1] − [Cd.9239]	1900 − 1918
4	[Cmd.1] − Cmd.9889	1919 − 1956
5	Cmnd.1 −	1956 −

(a)   On United Kingdom patent law

Report from the Select Committee on the Law relating to Patents for Inventions. House of Commons, 12 June 1829 (P.P 1829, Vol. III).

Report and Minutes of Evidence taken before the Select Committee of the House of Lords appointed to consider of the Bills for the further amendment of the Law touching Letters Patent for Invention. (P.P. 1851, vol. XVIII).

Report from the select Committee on the Patent Office, Library and Museum. House of Commons, 1864.

Report of the Commissioners appointed to 'inquire into the working of the Law relating to Letters Patent for Inventions' (Chairman: Lord Stanley) 29 Sept. 1864, London, 1865.

Report from the select Committee on Letters Patent. House of Commons, Vol. 1, 20 July 1871. Vol. 2, 8 May 1872.

Report of the Committee appointed by the Board of Trade to 'inquire into the duties, organization, and arrangements of the Patent Office under the Patents, etc, Act, 1883'. (Chairman: Lord Herschell) London, 1887.

Report of the Committee appointed by the Board of Trade to 'inquire into the working of the Patents Acts on certain specified questions'.
(Chairman: Sir Edward Fry)   Cd.   506, H.M.S.O., London, 1901
                Appendices   Cd.   530, H.M.S.O., London, 1901
Note by Solicitor-General on
            the Report   Cd. 1030, H.M.S.O., London, 1902

Proposed Amendments to the Patents and Designs Act, 1907.
Report of committee (Chairman: Parker of Waddington) 31st Oct. 1916

Royal Commission on Awards to Inventors.

First	Report,	Cmd.	1112,	H.M.S.O.,	London,	1921
Second	Report,	Cmd.	1782,	H.M.S.O.,	London,	1922
Third	Report,	Cmd.	2275,	H.M.S.O.,	London,	1925
Fourth	Report,	Cmd.	2656,	H.M.S.O.,	London,	1926
Fifth	Report,	Cmd.	3044,	H.M.S.O.,	London,	1928
Sixth	Report,	Cmd.	3957,	H.M.S.O.,	London,	1931
Final	Report,	Cmd.	5594,	H.M.S.O.,	London,	1937

Report of the Committee appointed by the Board of Trade in 1926 to consider the Dating of Patents, London 1927

Report of the Departmental Committee on the Patents and Designs Acts and Practice of the Patent Office (Chairman: Sir Charles Henry Sargent) Board of Trade, Cmd. 3829, H.M.S.O., London 1931

Minutes of Evidence taken before the above Committee; H.M.S.O., London, 1931

Patents and Designs Acts: First Interim Report of the Departmental Committee (Chairman: Kenneth R. Swan), Board of Trade, Cmd. 6618, H.M.S.O., London 1945

Patents and Designs Acts: Second Interim Report of the Departmental Committee; Board of Trade, Cmd. 6789, H.M.S.O., London, 1946

Patents and Designs Acts: Final Report of the Departmental Committee; Board of Trade, Cmd. 7206, H.M.S.O., London, 1947

Royal Commission on Awards to Inventors

First	report,	Cmd.	7586,	H.M.S.O.,	London,	1948
Second	report,	Cmd.	7832,	H.M.S.O.,	London,	1949
Third	report,	Cmd.	8743,	H.M.S.O.,	London,	1953

Crown Use of Unpatented Inventions; (Sir Harold Howitt committee) Cmd. 9788, H.M.S.O., London, 1956

The British Patent System; Report of the Committee to Examine the Patent System and Patent Law (Chairman : M.A.L. Banks, Esq.) Board of Trade, Cmnd. 4407, H.M.S.O., London, 1970

Manual of Office Practice (Patents) (loose leaf – periodically updated) 1st ed. 1968, 2nd ed. 1975 The Patent Office, London

Patent Law Reform: a consultative document. Dept. of Trade, H.M.S.O., London, 1975

Patent Law Reform: (White Paper) Cmnd. 6000, H.M.S.O., London, 1975

(b)   On international agreements

'International Convention for the Protection of Industrial Property'.

Paris,	20 Mar. 1883;	'Commercial	No	18	(1884)'	C. 4043
Brussels,	14 Dec. 1900;	'Treaty Series	No	15	(1902)'	Cd. 1084
Washington,	2 June 1911;	'Treaty Series	No	8	(1913)'	Cd. 6805
The Hague,	6 Nov. 1925;	'Treaty Series	No	16	(1928)'	Cmd. 3167
London,	2 June 1934:	'Treaty Series	No	55	(1938)'	Cmd. 5833
Lisbon,	31 Oct. 1958;	'Treaty Series	No	38	(1962)'	Cmnd. 1715
Stockholm,	14 July 1967;	'Treaty Series	No	61	(1970)'	Cmnd. 4431

'Convention on International Exhibitions'

| Paris, | 22 Nov. 1928; | 'Treaty Series | No | 9 | (1931)' | Cmd. | 3776 |
| | 10 May 1948; | 'Treaty Series | No | 57 | (1951)' | Cmd. | 8311 |

'European Convention relating to the Formalities required for Patent Applications'

| Paris, | 11 Dec. 1953; | 'Treaty Series | No | 43 | (1955)' | Cmd. | 9526 |

'Council of Europe Convention on the Unification of Certain Points of Substantive Law on Patents for Invention'

| Strasbourg, | 27 Nov. 1963; | 'Treaty Series | No | 47 | (1964)' | Cmd. | 2362 |

'Agreement concerning the Establishment of an International Patents Bureau'
The Hague,   6 June 1947;   Cmnd. 2679, 1965

'United Kingdom Patent Law: the Effects of the Strasbourg Convention of 1963';
Report by the Patents Liaison Group   Cmnd. 2835 1965

'Patent Cooperation Treaty'
Washington,   19 June 1907;   'Miscellaneous Series No 24 (1970)'   Cmnd. 4530

'Convention on the Grant of European Patents (European Patent Convention) with related documents';
Munich,   5 Oct.   1973;   'Miscellaneous Series   No   24   (1974)'

'Convention for the European Patent for the Common Market (Community Patent Convention)'
Luxembourg, 15 Dec. 1975: 'European Communities No 18 (1976), Cmnd. 6553

(all published by H.M.S.O., London)

104

REGULAR PUBLICATIONS

(i)   Weekly

The Commissioners of Patents' Journal Pub. at The Great Seal Patent Office
No. 1, 7 Jan. 1854 to No. 3130, 31 Dec. 1883

The Official Journal of the Patent Office No. 1, 4 Jan. 1844 to No. 469,
2 Jan. 1889

The Illustrated Official Journal (Patents) No. 1, 9 Jan. 1889 to No. 2195,
11 Feb. 1931

The Official Journal (Patents) No. 2196, 18 Feb 18 1931 to date

(ii)   Annual

Annual Report of the Commissioners of Patents for Inventions 1852/3, (Pub.
1854) to 1883 (pub. 1884).

Report of the Comptroller-General of Patents, Designs and Trade Marks
No. 1, 1883, pub. 1884, to date

Occasional Publications

The Patent Office has from its earliest days published leaflets of information
helpful to patent applicants, patentees and others. Current titles include:

Applying for a Patent, Revised April 1978

How to Apply for a Patent (from 1 June 1978)

4 BOOKS ON PATENT LAW AND PRACTICE AND THE HISTORY OF THOSE SUBJECTS

(a) Text books and monographs.

This section is divided into two parts. Part (i) lists works concerned with all the former United Kingdom patent law and practice and part (ii) lists works on the new United Kingdom and European patent systems which came into operation on 1 June 1978. United Kingdom patent applications filed, and patents granted, before that date are governed by the 1949 Act (and relevant rules) amended by the 1977 Act and 1978 Rules. Provided that the amendments are borne in mind, therefore, works relating to the 1949 Act are still valuable.

(i) On former United Kingdom law and practice.

The classic text books on patent law under the 1949 Act are those of Terrell and Blanco White. Both of these contain historical information; Terrell in the chapter introductions and Blanco White in the appendices. Books intended for a wider readership were those of Grace, Lees, Liebesny and Meinhardt. That by Lees was particularly suitable for inventors because it illustrated by means of a practical, albeit imagined, example, the procedure for patenting an invention. The authoritative sources of information on patent practice under the 1949 Act are the books prepared by the Chartered Institute of Patent Agents, and the loose-leaf manual published by the Patent Office. A book giving a broad and practical view of the whole topic of protecting industrial property, and so dealing with the registration of designs and trade marks and with copyright, as well as with patents, is that by Blanco White and Jacob. The most recent edition of this book, to which a third author has contributed, is listed in part (ii).

No book devoted exclusively to the history of the patent system seems to be readily available at present. Two classic post-war booklets were that of Gomme, mainly on the early history, and that of Harding, commemorating the Patent Office centenary. The study of the administration of the patent system by Boehm and Silberston contains an excellent historical chapter. Also the book by Baker on the patents for well-known inventions (see Section 5 of this bibliography) contains an interesting historical introduction.

A number of papers and other contributions to patents history have been published in the Transactions of the Chartered Institute of Patent Agents, and in their Journal C I P A, which replaced the Transactions in October 1971. References to some of these are given in the Institute books already mentioned.

Abbey, A	*Patent Protection*	Manchester, 1920
Abel, C D	*The Action of the Patent Laws in Promoting Invention* Taylor & Francis,	London, 1864
Abel, C D and Imray, J	*Patents, Designs and Trade Marks* Abel & Imray,	London, 1886
Agnew, W F	*The Law and Practice Relating to Letters Patent for Inventions* Wildy,	London, 1874
Association of the British Pharmaceutical Industry.	*Medicines and the Patent System*	London, 1972
Aston, J J	*The Law of Patents, Designs and Trade Marks* Waterlow & Layton	London, 1883
Aston, T	*The New Patents, Designs, and Trade Marks Act, 1883* Stevens,	London, 1884
Baraclough, W H	*Every Inventor his own Patent Agent* Wilson,	London, 1928
Bayly, J P	*Universal Information for Inventors*	London 1897. 1902
Bewes, W A	*Copyright, Patents, Designs, Trade Marks etc. (Manual of Practical Law)* Black,	London, 1891
Billing, S and Prince, A	*The Law and Practice of Patents and Registration of Designs* Benning,	London, 1845
Bliss, H J W	*British Patents and Designs Statutes as amended and consolidated to 1932* Maxwell,	London, 1932
Boehm, K (with Silberston, A)	*The British Patent System. 1: Administration* University of Cambridge Dept. of Applied Economics. Monograph 13. Cambridge University Press,	1967
Bosshardt, F, & Co. (pub)	*Notes on Inventions, Patents, Designs and Trade Marks* Manchester,	1915
Bousfield, W M	*The Patents, Designs and Trade Marks Act, 1883* Sweet,	London, 1884

Bradley, F E and Bowman, F H

The Inventor's Handbook of Patent Law and Practice
Seymour,   1914

Brewer, J A

Notes on the British Patents Acts
Author,   Glasgow, 1907

Brice, S W

The Law Practice and Procedure relating to Patents,
Designs and Trade Marks
Clowes,   London, 1885

Brown, J

Popular Treatise on the Patent Laws
Spon,   London, 1874

Browne, T B, Ltd (pub)

An Epitome of Useful Information relating to Trade
Marks, Letters Patent, Designs, Copyright and the Use
of the Royal Arms,
1st ed. 1903, 2nd ed. 1915   London,

Burke, P

Patent Law Amendment Act, 1852
Benning,   London, 1852
A compendium of the Patent Law as now amended
Benning,   London, 1857

Byrne, J P

A Handy Book of the Law and Practice of Patents for
Inventions and Registration of Designs
1st ed.   2nd ed. 1860 Davis

Campin F W

The Acts 28 Vict. cap.3 and cap.6, concerning
Inventions and Designs at Industrial Exhibitions etc.
Stevens & Haynes,   London, 1865
The Law of Patents for Inventions, with explanatory
notes on the Law of Designs and Trade Marks
Virtue,   London, 1869

Capsey, S R

Patents: an Introduction for Engineers and Scientists
Newnes-Butterworths,   1973

Carpmael, W

The Law of Patents for Inventions

1st ed.	Simpkin,	Marshall,	&	Weale 1832
2nd ed.	Simpkin,	Marshall,	&	Weale 1836
3rd ed.	Simpkin,	Marshall,	&	Weale 1842
4th ed.	Simpkin,	Marshall,	&	Weale 1846
5th ed.	Simpkin,	Marshall,	&	Weale 1852
6th ed.	Stevens,	Marshall,	&	Weale 1860

Carr, L H and Wood J C

Patents for Engineers
Chapman & Hall,   London, 1959

Chartered Institute of Patent Agents
     The Patents Act 1949
     Sweet & Maxwell,   London, 1950
     The Patents Acts 1949 to 1961
     Sweet & Maxwell,   London, 1968
     Annual supplements 1968 – 1974
     Patent Law of the United Kingdom
     Annual supplements 1975 –
     Sweet & Maxwell,   London, 1975

Clark, A M and Clark, W   *Analytical summaries of the Patents, Designs and Trade Marks Act, 1883*
     A M & W Clark,   London, 1884
     Patents for Inventions
     A M & W Clark,   London, 1889

Collier, J D   *An Essay on the Law of Patents for new Inventions,*
     Longmans & Rees,   London, 1803

Coryton, J   *A Treatise on the Law of Letters Patent for the sole use of Inventions*
     Sweet,   London, 1855

Coventry, C   *How to take out a Patent*
            Liverpool, 1901

Craig, A   *Patents, Trade Marks and Designs*
     The Bazaar,   London, 1879

Crossley, C W   *All about Patents*
            Guildford, 1911

Cruickshank and Fairweather Ltd. (pub.)
     The Law of Patents, Designs and Trade Marks
            Glasgow, 1908

Cunynghame, H.   *English Patent Practice*
     Clowes,   London, 1894

Daniel, E M   *A complete Treatise upon the new Law of Patents, Designs and Trade Marks*
     Stevens & Haynes,   London, 1884

Danson, J T and Drysdale, G
     The Inventors Manual
     Weale,   London, 1843

Davies, G   *Self-help to Patent Law*
     1870 (3rd ed.), 1895, 1906   Manchester

Davies, G and Prescott, T
>   The same.   1912   Manchester

Davies, G, Hunt, B and Hunt, E
>   *Handbook for Inventors and Patentees*
>   1859

Douglas, R   *Law for Technologists*
>   Gee,   London,   1964

Edmunds, L H   *The Patents, Designs and Trade Marks Acts (1883
>   to 1888) consolidated*   London,   1889

Edmunds, L (H) and Renton, A W   *The Law and Practice of Letters Patent
>   for Inventions*   1st   ed.
>   Stevens,   London,   1890
>   2nd   ed. (by Stevens, T M)
>   Stevens,   London,   1897

Edwards, F   *On Letters Patent for Inventions*
>   Hardwicke,   London,   1865

Ellis, R   *The Patent Law in relation to Chemistry*
>   Gill & Ellis   1912

Emery, G F   *Handy Guide to Patent Law and Practice*
>   Wilson   1st   ed.   London,   1896
>   Wilson   2nd   ed.   London,   1904
>   *The Solicitors' Patent Practice*
>   Wilson,   London,   1909

Evans-Jackson, J E   *Notes on Patents and Trade Marks*
>   Truscott,   1904

Fairbrother, H   *Fairbrother on Patents*   1913

Fairweather, W   *Patents for Inventions*
>   Glasgow,   1894

Flack, A M   *Patents, Designs and Trademarks*
>   Cassell's Work Handbooks
>   Cassell,   London,   1920

Fox, H G   *Monopolies and Patents: a study of the History and
>   Future of the Patent Monopoly*
>   Univ. of Toronto,   1947

Frank, W F   *The New Industrial Law*
>   London,   1950

110

Freeman, W M		*Patents and Designs Act, 1907*				
		Cox,				1908
Friswell, R J		*The Patents Bill, 1907*				
		Vacher,				1907
Frost, R		*A Treatise on the Law and Practice relating to Letters Patent for Inventions*				
		Stevens & Haynes,	1st ed.	London,		1891
		Stevens & Haynes,	2nd ed.	London,		1898
(2 vols.)		Stevens & Haynes,	3rd ed.	London,		1906
(2 vols.)		Stevens & Haynes,	4th ed.	London,		1912
		The Patents and Designs Act, 1907				
		Stevens & Haynes,		London,		1908

Fulton, D — *A Practical Treatise on Patents, Trade Marks and Designs*
London, 1894

2nd ed. *The Law and Practice relating to Patents, Trade Marks and Designs*
Jordan, London, 1902

3rd ed. *The Law and Practice relating to Patents, Trade Marks and Designs*
Jordan, London, 1905

4th ed. *The Law and Practice relating to Patents and Designs*
Jordan, London, 1910
The Patents and Designs Act, 1907
London, 1908

Gee, H T P — *Patents, Trade Marks and Designs*
1st ed. Gee & Co. Leicester, 1930
2nd ed. London, 1936

Gill, H A and Ellis, R — *Patents and their Exploitation*
Solicitors' Law S S London, 1912

Godfrey, R — *ABC of Patents*
Mason, Havant, 1972

Godson, R — *A Practical Treatise on the Law of Patents for Inventions and of Copyright*
1st ed. Butterworth London, 1823
1st supp. 1832, 2nd supp. 1835
2nd ed. Saunders & Benning, 1840
2nd ed. with supp. 1844, supp. 1851

Gomme, A — *Patents of Invention: origin and growth of the Patent System in Britain*
British Council, Longmans Green,
London, 1946

| Goodeve, T M | *Patent Practice before the Comptroller and the Law Officers* | | |
| | Sweet & Maxwell, | London, | 1889 |

Gordon, J W	*Monopolies by Patents and the Statutable Remedies Available to the Public*		
	Stevens,	London,	1897
	Compulsory Licences under the Patent Acts		
	Stevens,	London,	1899
	The Statute Law relating to Patents of Invention and Registration of Designs		
	Jordan,	London,	1908

| Grace, H W | *A Handbook on Patents* | | |
| | Knight, | London, | 1971 |

| Graham, J P | *Awards to Inventors* | | |
| | Sweet & Maxwell, | London, | 1946 |

| Gridley, H A A | *A Digest of Patent Law and Cases, incorporating the Provisions of the Patents Act, 1883* | | |
| | Ward, | London, | 1884 |

| Griffiths, A W | *Patent Law and Practice* | | |
| | Stevens, | London, | 1928 |

Haddan, R	*The Inventor's Adviser and Manufacturer's Handbook to Patents, Designs and Trade Marks*		
	1st ed. Harrison,	London,	1894
	2nd ed. Harrison,	London,	
	3rd ed. Harrison,	London,	
	4th ed. Harrison,	London,	1899
	5th ed. Harrison,	London,	1900
	6th ed. Harrison,	London,	1905
	7th ed. Harrison,	London,	1908
	8th ed. Harrison,	London,	1911
	9th ed. Harrison,	London,	1913
	10th ed. Harrison,	London,	1917
	11th ed. Harrison,	London,	1922
	12th ed. Harrison,	London,	1924
	A Compendium on Patents and Designs Law and Practice		
	Loose Leaf Harrison,	London,	1931 on
	Patents for Inventions: a concise guide for inventors and patentees		
	Pitman,	London,	1937

| Haes, H | *British Patent Law and Patentees' wrongs and rights* | | |
| | Whittingham, | London, | 1896 |

Handford, T J	*Notes on the Law and Practice relating to Letters Patent for Inventions*		
	Office for Patents,	London,	1882
Hands, W	*The Law and Practice of Patents for Inventions*		
	Clarke,	London,	1808
Harding, H	*Patent Office Centenary*		
	H.M.S.O.	London,	1953
Hardingham, G G M	*Patents for Inventions and How to Procure Them*		
	Lockwood,	London,	1891,1902
	Patent Rights Lockwood,	London,	1908
Harrison, J	*Abstract of the 15 & 16 Vict. (Patent Law Amendment Act, 1852) c.83 with suggestions of Inventors, etc.*		
		Birmingham,	1852
Hart, W F	*Information on the Patent Law of Great Britain, Ireland, and the Isle of Man*		
	Leather Trades,	London,	1888

Haseltine, Lake & Co. (pub.) *Information and notes on some of the provisions of the Patents and Designs Act, 1919 and the Trade Marks Act, 1919*   London,   1920
An Outline of the Law of Letters Patent for Inventions in the United Kingdom   London,   1929

Hemming, H B	*Practical Guide to the Law of Patents*		
	Waterlow,	London,	1905
Henry, M	*A Defence of the present Patent Law*		
	Office for Patents,	London,	1866
Higgins, C	*A Concise Treatise on the Law and Practice of Patents for Inventions*		
	Clowes,	London,	1884
Hindmarch, W M	*A Treatise on the Law relating to Patent Privileges for the Sole Use of Inventions*		
	Stevens; and Norton; and Benning London,		1846
Hollins, C	*Introduction to the Patenting of Inventions*		
	Benn,	London,	1951
Holroyd, E	*A Practical Treatise of the Law of Patents for Inventions*		
	Richards,	London,	1830

Houghton, B	*Technical Information Sources: a guide to Patents, Standards, and Technical Reports Literature*
	1st ed. 1967 2nd ed. 1972
	Bingley, London,

| Hulme, E W | *Early History of the English Patent System* |
| | Little & Brown, Boston, U.S.A. 1909 |

| Hunt, B | *Handbook to the Patents, Designs and Trade Marks Act, 1883* |
| | Waterlow, London, 1884 |

Johnson, J
The Patentee's Manual: being a treatise on the law and practice of letters patent

| 1st | ed. | Longman, Brown etc. | London, | 1853 |
| 2nd | ed. | Longman, Brown etc. | London, | 1858 |

Johnson, J and Johnson, J H

3rd	ed.	Longman, Green etc.	London,	1866
4th	ed.	Longmans Green and Stevens, London,1879
5th	ed.	Longmans Green and Stevens, London.1884
6th	ed.	Longmans Green and Stevens, London,1890

The Patents, Designs and Trade Marks Act, 1883

1st	ed.	Longmans Green, and Stevens,
		London,	1883
2nd	ed.	Longmans Green, and Stevens,
		London,	1884

An Epitome of Law and Practice connected with Patents for Inventors

1st	ed.	Longmans Green and Stevens,
		London,	1887
2nd	ed.	Longmans Green and Stevens,
		London,	1894
3rd	ed.	Longmans Green and Stevens,
		London,	1900

| Keith J | *Our Patent Laws* | London, |

Kings Patent Agency Ltd. (pub)   *Patents and Trade Marks*
Booklet — numerous editions (21st ed. 1950)

Lawson, W N
The Practice as to Letters Patent for Inventions . . . under the Patents, Designs and Trade Marks Act, 1883

| 1st | ed. | Butterworths, | London, | 1884 |
| 2nd | ed. | Butterworths, | London, | 1889 |

assisted by Sharp, C and Warmington, M D

| 3rd | ed. | Butterworths, | London, | 1898 |

114

Lees, C	*Patent Protection: the inventor and his patent*		
	Business Pub.	London,	1965
Leverson, M R	*Copyright and Patents*		
	Wildy,	London,	1854
Liebesny, F (ed.)	*Mainly on Patents: the use of industrial property and its literature*		
	Butterworths,	London,	1972
Linley, C M	*Practical Advice to Inventors and Patentees*		
	Pitman,	London,	1925
Lochner, R	*The New Patents Act*		
		London,	1951
Lund, H	*A Treatise on the Substantive Law relating to Letters Patent for Inventions*		
	Sweet,	London,	1851
Lunge, E	*Compulsory Working and Revocation of Patents*		
	Stevens,	London,	1910
Macfie, R A	*The Patent Bills of 1883*		
	Clark,	Edinburgh,	1883
	Copyright and Patents for Inventions		
	Vol. 1, 1879 Vol. 2, 1883 Clark, Edinburgh		
Macgregor, J	*The Language of Specifications of Letters Patent for Inventions*		
	Benning,	London,	1856
Macpherson, W	*The Practice of the Judicial Committee of Her Majesty's Most Honourable Privy Council*		
	Sweet,	London,	1860
Marks, G C	*Inventions, Patents and Designs*		
	Technical Publishing,	London,	1907
	Notes and Judgments on the "Working" of British Patents		
	Technical Publishing,	London,	1910
	Inventions and Industries under the "Working" Clauses of British Patent Acts		
	Technical Records,	London,	1922
	The Patents and Designs Acts		
	Sweet & Maxwell,	London,	1933

Martin, W	*The Construction or Interpretation of Specifications of Letters Patent*		
	Kelly Law Book Co,		1900
	The English Patent System		
	Dent,	London,	1904
Meinhardt, P	*Inventions, Patents and Monopoly*		
	1st ed. Stevens,	London,	1946
	2nd ed. Stevens,	London,	1950
	Inventions, Patents and Trade Marks		
	Gower Press,	London,	1971
Miller, D B	*How to Patent and Commercialize your Inventions*		
	Pitman,	London,	1937

Model Engineer Series No. 20. *Patents Simply Explained*   1904

Morgan, T W	*Compulsory Licences and Revocation of Patents*		
	Solicitors' Law S S,	London,	1913

Moritz, R and Hereward, W H   *Post-War Patent Practice*
       Sweet & Maxwell,   London,   1921

Moulton, H F	*The Present Law and Practice relating to Letters Patent for Inventions*		
	Butterworths,	London,	1913

Moulton, H F and Evans-Jackson, J H   *The Patents, Designs and Trade Marks Acts*
       1st ed. Butterworths,   London,   1920
       2nd ed. Butterworths,   London,   1930

Munden, W J	*How to take out your own Patents*	
		London, 1894, 1896
Munro, J E C	*The Patents Acts, 1883*	
		London, 1884
Neilson, R M	*Handbook on Patents for Scottish Solicitors*	
		Glasgow, 1914
Newby, F	*How to Find out about Patents*	
	Pergamon,	London, 1967

Newton, A V	*Patent Law and Practice*			
	1st ed. (anonymous)	Trübner,	London,	1871
	2nd ed.	Trübner,	London,	1879
	3rd ed.	Cox,	London,	1893
	4th ed.	Cox,	London,	1897
	An Analysis of the Patents, Designs and Trade Marks Act, 1883			
			London,	1883

	An Analysis of the Patent and Copyright Laws London, 1884
Nicolas, V	*The Law and Practice relating to Letters Patent for Inventions* Butterworths, London, 1904 *The Patents Act, 1902, and Notes thereon* Butterworths, London, 1904
Norman, J P	*A Treatise on the Law and Practice relating to Letters Patent for Inventions* Butterworths, London, 1853
O'Brien, J O	*British Patents. Inventor's Pocket Book* Manchester, 1914
Parsons, C S	*Patents, Designs and Trade Marks* Technical Press, London, 1938
Phillips, R E	*One Thousand Patent Facts* Iliffe, London, 1895
Phillips, R E and Flack, A M	*The ABC Guide to Patents for Inventions* Butterworths, London, 1915
Price, W H	*The English Patents of Monopoly* Constable, London, 1906 Reprint, 1976
Prideau, F	*Patents Conveyancing* Stevens, London, 1927
Prosser, R	*Patent Law Amendment Act, 1852* Birmingham, 1852
Rankin, R R	*An Analysis of the Law of Patents* London, 1824
Roberts, J	*The Grant and Validity of British Patents for Inventions* Murray, London, 1903 *The Inventor's Guide to Patent Law and the New Practice Practice* Murray, London, 1905

	Notes on the New Practice at the Patent Office		
	Eyre & Spottiswoode,	London,	1906
Roberts, J and Moulton,	The Patents and Designs Act, 1907		
H F	Butterworths,	London,	1907

Romer, C	Practice before the Comptroller of Patents		
	1st ed. Sweet & Maxwell,	London,	1911
	2nd ed. Sweet & Maxwell,	London,	1926

Royston, E R, & Co	Practical Guide to Patents for Inventions		
		Liverpool,	1902

Rushen, P C	A Critical Study of the Form of Letters Patent for Inventions		
	Stevens,	London,	1908

Sime, J and Thomson, R C	The Inventor's Guide		
	Thomson & Co	Glasgow,	1901,1906

Smith, J W	An Epitome of the Law relating to Patents for Inventions		
	1st ed. Maxwell,	London,	1836
	2nd ed. Maxwell,	London,	1854

Smith, T E	Inventions and how to Patent them	London	1890

Spence, W	A Treatise on the Principles relating to the Specification of a Patent for Invention		
	Stevens & Norton,	London,	1847
	Patentable Invention and Scientific Evidence		
	Stevens & Norton,	London,	1851
	Practical Remarks on the Present State of the Law of Patents		
	1st ed. Stevens & Norton,	London,	1852
	2nd ed. Stevens & Norton,	London,	1856
	Look to your Provisional Specifications		
	1st ed.	London,	1857
	2nd ed.	London,	1861
	Specifications: a word of advice to Patentees		
		London,	1866
	Specifications as bases of Patents		
	1st ed. 1870 2nd ed. 1875	4th ed.	1884

Stanley, A H	Patents to Inventors		
	Marshall,		1906

Swan, K R	*The Law and Commercial Usage of Patents Designs and Trade Marks*		
	Constable,	London,	1908

Taylor, W H	*Patents: how to obtain them*		
		Manchester,	1901
	The Inventor's and Patentee's Year Book		
	1st ed. Dexter,	Manchester,	1914
	2nd ed. Dexter,	Manchester,	1924

Terrell, T	*The Law and Practice relating to Letters Patent for Inventions* Sweet & Maxwell,	London,	
	1st ed.		1884
	2nd ed.		1889
	3rd ed. (by W P Rylands)		1895
	4th ed. (by C Terrell)		1906
	5th ed. (by C Terrell)		1909
	6th ed. (by C Terrell and A D Jaffé)		1921
	7th ed. (by C Terrell and D H Corsellis)		1927
	8th ed. (by J R Jones)		1934
	9th ed. (by K E Shelley)		1951
	10th ed. (by K E Shelley)		1961
	11th ed. (by G Aldous, D Falconer and W Aldous)		1965
	12th ed. (by D Falconer, W Aldous and D Young)		1971

Thomas, C W	*Patents, Trade Marks and Designs* (Nutshell Series).		
	Ocean Publishing,	London,	1932

Thompson, W P	*Handbook of British Patent Law*		
	Stevens,	London,	1892,1912

Thomson, W R M	*The Inventor's Guide*		
		Glasgow,	1897

Thornton, A A	*Thornton on Patents* (British and Foreign)		
	Jones,	London,	1910

Turner, T	*Counsel to Inventors of Improvements in the Useful Arts*	London,	1850
	The Law of Patents and Registration of Invention and Designs		
		London,	1851

Waggett, J F	*Law relating to the Prolongation of the term of Letters Patents for Inventions*		
	Butterworths,	London,	1887

Walker, J E and Foster, R B *Patents for Inventions*
                 1st   ed.   Pitmans   London,   1922
                 2nd   ed.   (by J E Walker and J Roscoe)
                                 Pitmans,   London,   1936

Wallace, R W *The Patents, Designs and Trade Marks Act, 1883*
                 Maxwell,   London,   1884

Wallace, R W and Williamson, J B *The Law and Practice relating to Letters Patent for Inventions*
                 Clowes,   London,   1900

Webster, T *The Law and Practice of Letters Patent for Inventions*
                 Crofts & Blenkarn,   London,   1841
               On the subject matter of Letters Patent for Inventions
               Supp. to The Law and Practice . . .
                 Crofts & Blenkarm,   London,   1841
               2nd ed. *On the subject-matter, title and Specification of Letter Patent for Inventions and Copyright of Designs*   1848
               3rd   ed.   *The subject-matter of Letters Patent for Inventions and Registration of Designs*
                 Elsworth,   London,   1851
               The New Patent Law: its history, objects and provisions
               1st   ed.   Elsworth,   London,   1852
               2nd   ed.   Elsworth,   London,   1852
               3rd   ed.   Elsworth,   London,   1853
               4th   ed.   Chapman & Hall; Elsworth
                                   London,   1854
               On Property in Designs and Inventions in the Arts and Manufactures, Chapman & Hall,   London,   1853

Wetherfield, T *Patent Law and Practice*
               1st   ed.   1842
               2nd   ed.
               3rd   ed.
               4th   ed.   1854

Wheatcroft, W H *Notes on the Law of Torts and Patents and Copyright*
                                  Cambridge,   1913

Wheeler, G J *Notes on the Prolongation of Letters Patent for Inventions*
                 Eyre & Spottiswoode,   London,   1898

White, T A Blanco *Patents and Registered Designs* (This is the Law)
                 Stevens,   London,   1947
               2nd   ed.   *Patents — Law and Legislation*
                                 (This is the Law)
                 Stevens,   London.   1950

Patents for Inventions and the Registration of Industrial Designs

1st	ed.	Stevens,	London,	1950
2nd	ed.	Stevens,	London,	1955
3rd	ed.	Stevens,	London,	1962
4th	ed.	Stevens,	London,	1974

Industrial Property and Copyright
Combining *This is the Law* volumes

Stevens,	London,	1962

White, T A Blanco and Jacob, R *Patents, Trade Marks, Copyright and Industrial Designs* (Concise College Texts)
(revision of *Industrial Property and Copyright*)

1st	ed.	Sweet & Maxwell,	London,	1970

Whitelow, E T *The English Patents Acts arranged as a Code*
London, 1903

Wildbore, H J W *Patents Explained*
London, 1933

Wise, W L *A Summary of the New Patent Act, 1883*
London, 1883

Withers, J S and Spooner *Guide to Patents, Trade Marks and Designs*
Withers & Spooner, 1914

Wordsworth, C F *A Summary of the Law of Patents for Inventions and of Extension of Patents*

1st	ed.	Benning,	London,	1853
2nd	ed.	Benning,	London,	1857

(ii)   On the new United Kingdom and European patent systems

Three of the listed works are in loose-leaf form so that as the law and practice under the 1977 Act and the European Patent Convention are developed, they can be brought up to date with additional pages. Information on the new patent systems is also given in recent works on international patenting listed in section (d) of this bibliography.

Bard, B (editor) *Applications and Limitations of the Patent System*
(Conference proceedings)
IPC Science & Technology Press,
Guildford 1975

Brett, H *The Patents Act 1977 - an introductory guide*
ESC Publishing, Uppingham 1977

The United Kingdom Patents Act 1977
2nd edition of *The Patents Act 1977*
ESC Publishing, Uppingham 1978

Chartered Institute of Patent Agents *European Patents Handbook*
2 vols. loose-leaf. Oyez, London, 1978

Pennington, R R (editor) *European Patents at the Crossroads*
(Conference proceedings) Oyez, London, 1976

Phillips, J *Employees' Inventions and the Patents Act 1977*
Mason, Havant, 1978

Vitoria, M (editor) *The Patents Act, 1977* Queen Mary College Patent
Conference Papers.
Sweet & Maxwell, London, 1978

Walton, A M and Laddie, H I L *Patent Law of Europe and the United Kingdom*
Loose-leaf. Butterworths, London, 1978

White, T A Blanco, Jacob, R, Jeffs, J and Cornish, W R *Encyclopedia of United
Kingdom and European Patent Law*
loose-leaf. Sweet & Maxwell and W Green,
London, 1978

White, T A Blanco, Jacob, R and Davies, J D *Patents, Trade Marks, Copyright
and Industrial Designs* (Concise College Texts)
2nd ed. of work listed in part (i) above.
Sweet & Maxwell, London, 1978

(b) Authors of leaflets.

Brewster, E H G	1894
Day, Davies and Hunt,	1886
Fairfax, J S	1896
Hughes and Young,	1898
Johnson, W and Johnson, J H	1851
Marsh, J	1921
Mathys, A W	1909
Radford, R H	1910
Reason, E H	1911
Tasker and Co.	1904
Whiteman, W T	1884

(c) Authors of works on patent law reform

Association of Patentees and Proprietors of Patents,	1851
Ayrton, S G	1890
Bennett, I	1855
Drewry, C S	1851, 1863

George, D L	1908
Grierson, F W	1881
Historicus (pseudonym)	1919
Hughes, E J	1851
Hunt, E	1859,1870
Lawson, J A	1851
Livesey, P	1876
Macfie, R A	1864,1869,1875
Mann, A	1865
Marks, G C	1907
Newall, R S	1848
(reprinted 1870)	
Newton, A V	1864
Polanyi, M	1945
Roberts, R	1830,1833
Soul, M A	1863,1869
Stanley, Lord	1856
Thorold, W	1829
Wallington, R A	1851
Wheatley and Mackenzie	1893
Wilson, R	1865
Woodcroft, B	1851

(d) Works on international patenting

Baxter, J W — *World Patent Law and Practice*
1st ed.   Sweet & Maxwell,   London,   1968
2nd ed.   Sweet & Maxwell,   London,   1973
Annual supplements were issued to each edition.

A loose-leaf version was published in the U.S.A. by
Bender, New York, as Vol. 2 (1968) of 'Patent Law
and Practice'.

Boughton, H F — *The Patentee's Guide*
Simpkin, Marshall etc.   London,   1890

Bougon, J — *The Inventor's Vade Mecum*
London,   1870

Boult, A J — *Digest of British and Foreign Patent Laws*
1st ed.   Bemrose,   London,   1895
2nd ed.   Boult, Wade & Kilburn, London,   1899

Boult and Tennant (pub) — *Summary of British and Foreign Patent Laws*
1908

Calvert, R (editor) — *The Encyclopedia of Patent Practice and Invention
Management*   Reinhold,   New York, U S A   1964

123

Carpmael, A and Carpmael, E   *Patent Laws of the World*
                             1st   ed.   Clowes,   London,   1885

	1st	ed.	Clowes,	London,	1885
	2nd	ed.	Clowes,	London,	1889
	Supplement		Clowes,	London,	1889

Chartered Institute of Patent Agents   *Patent Laws of the World*

	3rd	ed. of Carpmael work	1899 on
	4th	ed.	1911 on

Chatwin, J   *The Protection of Industrial Property*
                     Chatwin & Co. Various editions.

Curtis, W J   *Information as to the mode of obtaining protection for Inventions in the British Colonies*   1857

Davies, J   *Pamphlet on Patents*
                   Weale; Simpkin,   London,   1850

Day, C A   *Handbook on British, Colonial and Foreign Patents*
                                   London,   1895

Derwent Publications Ltd. (pub)   *Derwent Patents Manual* London,   1962

Edwards, E and Edwards, A E   *How to take out Patents in England and Abroad*

	1st	ed.		London,	1905
	2nd	ed.	(A E Edwards only) Wyman,	London,	1912

Fraser, J   *Handybook of Patent and Copyright Law − English and Foreign*
                   Sampson Low,   London,   1860

Goold., L W (pub)   *Pamphlet on British, Foreign and Colonial Patents*
                                 Birmingham,   1908

Hart, W F   *Tabular Statement of the Patent Laws of the World*
                                 London,   1890

Haseltine & Lake (pub)   *Tables of Information relating to European patents*
                                 London,   1887

Hughes, E J   *The Patent Laws of All Nations, and the Acts for the Registration of Designs*
                                Manchester,   1854

Justice, P M   *Epitome of British, Foreign and Colonial Patent Laws*
                                   London,   1884

Katzaroff (or Katzarov), K Patent Directory
1924 to 1940? Sofia, Bulgaria
Later editions, Geneva, Switzerland
Latest (8th) edition (in two volumes)   1976

Kilburn, B E D   *Summary of British and Foreign Patent Laws*
Kilburn and Strode,   1914

Leechman, G D   *British, Colonial and Foreign Patents*
London,   1897

Loosey, C F   *Collection of the Laws of Patent Privileges of Europe,*
United States and Dutch West Indies,   1849

Murdock, H H   *Information respecting British and Foreign Patents and*
the Registration of Designs Office of Patents

O'Brien, D   *Inventive Property Guide*
Lithocraft,   Liverpool,   1970

Octrooibureau los en Stigter, Amsterdam, Holland
Manual for the Handling of Applications for Patents,
Designs and Trade Marks throughout the World
1st ed. 1927. Loose-leaf since 1936

Schade, H   *Patents at a glance: a survey of substantive law and*
formalities in 46 countries
2nd ed. Heymann,   Munich   1971

Thompson, W.P.   *Handbook of Patent Law, British and Foreign*
Liverpool,   1874
8th ed. (with A J Boult), Stevens London,   1889
Numerous editions, at least to 1920

Tolhausen, A   *A Synopsis of the Patent Laws of Various Countries*
1st ed.   1857
2nd ed. Trübner,   London,   1869

White, W W and Ravenscroft, B G   *Patents throughout the World*
1st ed. frequent revisions to 1977 (loose-leaf)
2nd ed. (ed. A N Greene) 1978 (loose-leaf 2 vols)

Urling, R W   *The Laws of Patents in Foreign Countries*   1845

Warden, J (editor)   *Annual of Industrial Property Law*
1975 Shepheard-Walwyn
1976 Common Law Reports
1977 European Law Centre

Wise, W L   *Gleaning from Patent Laws of all Countries*
                                                    London,   1895

5 SELECTED BOOKS ON INVENTIONS, THEIR HISTORY AND EXPLOITATION

Baker, R   *New and Improved: inventors and inventions that have*
                            changed the modern world
                            British Museum Pubs,   London,   1976

Hope, A   *Why didn't I think of it first?*
                            David & Charles,   Newton Abbot

Jewkes, J, Sawers, D and Stillerman, R   *The Sources of Invention*
                            1st ed. 1958 Revised 1962   2nd ed. 1969
                            Macmillan,   London

Taylor, C T and Silberston, Z A *The Economic Impact of the Patent System:*
                            a study of the British Experience
                            University of Cambridge Dept. of Applied Economics.
                            Monograph 23.
                                Cambridge University Press   1973
                            (Part 2 of *The British Patent System*; see Boehm in part
                            4 (a) (i)) of this bibliography)

Thring, M W and Laithwaite, E R   *How to Invent*
                            Macmillan,   London,   1977

INDEX

INDEX

129

135